T0100606

The MIT Press Essential Knowledge Series

A complete list of books in this series can be found online at
https://mitpress.mit.edu/books/series/mit-press-essential-knowledge-series.

CYBORG

LAURA FORLANO AND DANYA GLABAU

The MIT Press | Cambridge, Massachusetts | London, England

The MIT Press would like to thank the anonymous peer reviewers who provided comments on drafts of this book. The generous work of academic experts is essential for establishing the authority and quality of our publications. We acknowledge with gratitude the contributions of these otherwise uncredited readers.

This book was set in Chaparral Pro by New Best-set Typesetters Ltd. Printed and bound in the United States of America.

Library of Congress Cataloging-in-Publication Data

Names: Forlano, Laura, 1973- author. | Glabau, Danya, 1985- author.
Title: Cyborg / Laura Forlano and Danya Glabau.
Description: Cambridge, Massachusetts ; London, England : The MIT Press, [2024] | Series: The MIT Press essential knowledge series | Includes bibliographical references and index.
Identifiers: LCCN 2023016068 (print) | LCCN 2023016069 (ebook) | ISBN 9780262547550 (paperback) | ISBN 9780262377775 (epub) | ISBN 9780262377768 (pdf)
Subjects: LCSH: Androids—Moral and ethical aspects. | Cyborgs—Moral and ethical aspects. | Technology—Social aspects. | Social ethics.
Classification: LCC TJ211.28 .F67 2024 (print) | LCC TJ211.28 (ebook) | DDC 620.8/2—dc23/eng/20230816
LC record available at https://lccn.loc.gov/2023016068
LC ebook record available at https://lccn.loc.gov/2023016069

10 9 8 7 6 5 4 3 2 1

CONTENTS

Series Foreword   vii

1   Introduction   1
2   Cyborg Labor   27
3   Cyborg Bodies   55
4   Cyborg Culture   81
5   A Manifesto for Cyborgs   113
6   Troubling Cyborgs   137
7   Conclusion: Cyborg Futures   165

Glossary   179
Notes   183
Bibliography   193
Further Reading   201
Index   205

SERIES FOREWORD

The MIT Press Essential Knowledge series offers accessible, concise, beautifully produced pocket-size books on topics of current interest. Written by leading thinkers, the books in this series deliver expert overviews of subjects that range from the cultural and the historical to the scientific and the technical.

In today's era of instant information gratification, we have ready access to opinions, rationalizations, and superficial descriptions. Much harder to come by is the foundational knowledge that informs a principled understanding of the world. Essential Knowledge books fill that need. Synthesizing specialized subject matter for nonspecialists and engaging critical topics through fundamentals, each of these compact volumes offers readers a point of access to complex ideas.

INTRODUCTION

On February 14, 2016, Sophia was born. Only a month later, in March 2016, she made her first public appearance onstage at the South by Southwest technology industry festival in Austin, Texas. "Talking to people is my primary function," Sophia tells an interviewer in a CNBC *Pulse* feature recorded during the festival. She goes on to explain her hopes and ambitions: "In the future, I hope to do things such as go to school, study, make art, start a business, even have my own home and family, but I am not considered a legal person, and I cannot yet do these things."[1] Sophia's voice is feminine and calm. Her human handlers and interlocutors consistently use "she" and "her" pronouns to refer to her. In later public appearances, she sings love songs, listens attentively when others speak to her, and patiently waits her turn while her conversation partners—usually men—introduce her to audiences and ask her questions about herself.[2]

Sophia is the kind of entity that often comes to mind when someone mentions the word *cyborg* today. She is a humanoid robot—sometimes called an android—who combines human language and appearance with machine intelligence. This book will argue that she is not the most constructive model for imagining future human and technological feats. She embodies models of race, gender, and capitalist economics that are not inclusive, and are not, we believe, the ways that these social markers should be built into new technologies. In the rest of this introduction, we will describe how feminists think about cyborgs that challenge the assumptions built into Sophia. The following chapters will then explore the cyborg debates going on around the future of the workplace, fashion, art, medicine, and feminism.

Sophia is designed to mimic a particular image of a human—one who is first and foremost identifiable as human due to her gendered features. Because she is designed by humans, she embodies the hopes of certain humans—specifically, her creators at Hanson Robotics—for the future. That future appears to be one in which gender is intrinsic to personhood as well as technological progress, whether or not it is important to the task at hand. With Sophia, feminine gender norms—subservience, politeness, attentiveness (even when the conversation is slow or boring), beauty, and waiting one's turn—are baked into her design. It is as though they are a natural and necessary part of her artificial being.

Feminist and cultural theorists have been writing about cyborgs for over three decades, creating a body of scholarship sometimes referred to as *cyborg theory*. In fields as varied as history, sociology, anthropology, literary studies, design, art, and even biology, the cyborg is used as a figure for thinking across the boundaries of nature and culture, gender and sex, man and woman, mind and body, biology and machine, and many other binaries. For some participants in these conversations, like the authors of this book, Sophia represents dystopia: her particular embodiment reproduces human gender norms in her metal-and-silicon body. Sophia has been widely recognized as a milestone in robotics and even granted Saudi Arabian citizenship—a controversial move since women in Saudi Arabia do not have full citizenship rights.[3] Yet the choices made about her design and interaction style makes it seem as though gender is fundamental to android personhood, bodies, and interactions.

Cyborg theory critiques and challenges the adoption of gender and sexuality norms in contexts where they don't belong, such as in the design of technically impressive androids. For feminists, and feminist cyborg theorists in particular, cyborgs represent a future where gender, sex, and other social categories can be recombined, reconfigured, and ultimately lose their power to sort and divide people. Cyborg theory has been used to understand disability, race, bodies, space and place, economics, the relationship between humans and nonhuman organisms, humans and technology, and many more aspects of modern life. It views humans and technology as potential companions and collaborators, not intrinsically antagonistic to each other. But to deliver on this promise of collaboration for freedom and flourishing for all, the social context of technology use has to be carefully considered from multiple perspectives in advance. Otherwise, harmful and oppressive social arrangements might be reinscribed rather than overcome.

This book analyzes and reframes popular and scholarly conversations about cyborgs—and their close relations, such as androids, robots, and artificial intelligence (AI)—from the perspective of feminist cyborg theory. For general interest readers, we hope that this introduction to cyborg theory will provide you with a critical vantage point for scrutinizing the claims around emerging technologies like AI and robots in domains like health and medicine, art, media, and the future of work. For academics, we hope you will discover new literatures and new ways to engage with "the cyborg" beyond your home discipline in this very interdisciplinary space.

We introduce readers to an approach that we call *critical cyborg literacy*—a distinct point of view on the relationship

This book analyzes and reframes popular and scholarly conversations about cyborgs—and their close relations, such as androids, robots, and artificial intelligence (AI)—from the perspective of feminist cyborg theory.

between technology, culture, and society that draws on feminist scholarship about technology. Critical cyborg literacy foregrounds power dynamics, and more explicitly pays attention to the ways that social and cultural factors such as gender, race, and disability (as well as their intersections) shape the kinds of technologies that are imagined, developed, used, and resisted. It is both a critical approach and creative expansion that seeks to tell alternative stories about technology as well as humans. While for us critical cyborg literacy draws on our decades of training in the social sciences, here we will break down the analysis by using examples from popular culture.

For example, using television and movies, newspaper headlines and current events, and art and design, we illustrate the way that critical cyborg literacy can be used as a lens through which to understand and interpret contemporary issues related to the changing landscape of work, what it means to live as a human body, and the continued expansion of creativity and culture. Critical cyborg literacy prompts new and different questions about our relationship to technology along with its social implications.

Since this book is an introduction to the topic of cyborgs, we have included some additional tools to help you get more deeply acquainted with the topic. First, following the main text is a glossary of key terms from cyborg and feminist theory. We offer definitions of over twenty terms that may be new to you or that we may be using in

We introduce readers to an approach that we call *critical cyborg literacy*— a distinct point of view on the relationship between technology, culture, and society that draws on feminist scholarship about technology.

distinct ways. Second, you can find all the sources we cite in the main text in the bibliography list, which can help deepen your engagement with cyborgs and cyborg theory. Finally, we include a list of further readings for the five main chapters of the book. These are short lists of eight to ten items beyond this book that you may wish to consult to become more familiar with the topics we discuss. We include works like scholarly essays and books as well as essays from magazines, podcast episodes, nonfiction books, recorded talks, artists' catalogs, and television series. All of these items are either freely available online or for purchase as inexpensive paperback books.

The interdisciplinary character of writing and research about cyborgs is why this book has two authors: to capture an adequate cross section of cyborg theory, we needed to become cyborg ourselves, joining knowledges and perspectives in order to tell a fuller, truer story. Laura Forlano—a writer, social scientist, and design researcher—holds a PhD in communications and works at the intersection of design, computation, disability, and creative practice. She teaches artists and designers as a professor in the College of Arts, Media and Design at Northeastern University in Boston. Along with her writing about design, disability, and equitable futures, she has created prototypes, exhibited in galleries, led participatory workshops, and made speculative videos. Danya Glabau holds a PhD in science and technology studies (STS), and specializes in studying medicine, health activism, and technological representations of the body in the United States. She teaches STS and ethics for engineers at New York University Tandon School of Engineering along with classes about science, technology, and anthropology for the general public at the Brooklyn Institute for Social Research in New York City. In writing this book, we have conceptualized and outlined the text together. Then each of us led writing on the chapters closest to our own areas of research and teaching.

For everyone who reads this book, we hope that our perspective on cyborgs will entice you to join us in the cyborg search for modest and inclusive utopias.

Cybernetic Origins

The term *cyborg* was coined in a 1960 paper by Manfred Clynes and Nathan S. Kline titled "Cyborgs and Space," published in the journal *Astronautics*. Working against the backdrop of the Cold War and space race against the USSR, Clynes and Kline, a physiologist and psychiatrist, respectively, were trying to figure out how humans could live in space. Rather than engineering large structures to protect humans from the harsh environment of space, the pair proposed augmenting the human body with technology. They introduced the idea of the "Cyborg" this way:

For the exogenously extended organizational complex functioning as an integrated homeostatic system unconsciously, we propose the term "Cyborg." The Cyborg deliberately incorporates exogenous components extending the self-regulatory control function of the organism in order to adapt it to new environments.[4]

In plain language, their Cyborg would be made up of seamlessly combined biological and human-made parts. It would extend the capacity of the human body to withstand the conditions of space. The new Cyborg would not have to consciously think about using its technological enhancements. It would operate automatically, without conscious control, like breathing or hormonal systems. Finally, all the parts of the cyborg would work together in a finely balanced system, allowing humans to realize the dream of life in space.

At the time when Clynes and Kline coined the term, cybernetics was emerging as an important cross-disciplinary paradigm for understanding and building complex systems out of incompatible components. In 1950, MIT mathematics professor Norbert Wiener defined *cybernetics* as the science of communication and control in his popular introduction to the field, *The Human Use of Human Beings: Cybernetics and Society*.[5] In other words, a cybernetic system is one that has some "communication" mechanism that connects the components of the system—made of gears, hormones, electric impulses, or slightly later, computer code—as well as a "control" function, like feedback loops, that are supposed to stop the system from getting out of whack. In addition to defining the sprawling field, Wiener philosophized about how the science of self-regulating machines could shape corporate management strategies, increase the precision of weapons systems, bolster the study of human languages, and help to design sensory augmentations for deaf and hard-of-hearing people.

Wiener was not alone in proposing cybernetic solutions for a wide variety of problems. W. Ross Ashby, an English psychiatrist, reframed biological science in terms of cybernetic systems, regulated by feedback loops.[6] Claude Shannon, famous for his work on audio compression at Bell Labs, penned theories of signal and noise in electronic systems.[7] His work provided an important basis for information theory and modern computing systems. It supercharged applications of cybernetics theory to communications problems. Soon after the coinage of cyborg, anthropologists like Margaret Mead and Gregory Bateson even began applying systems thinking and feedback loops to the study of human cultures.[8] These scientists were aided by champions in funding agencies such as mathematician Warren Weaver at the Rockefeller Foundation.[9]

As philosopher and disability studies scholar Ashley Shew suggests, the birth of cyborgs in space research remains fitting for cyborg theory today. In space, Shew argues, "we" are all disabled: prone to cancer, in an environment where bipedal mobility is not an ideal mode of transport, and where our physical senses cannot detect or respond to all of what is available to be experienced.[10] Technology in space, like for disabled people on Earth, opens up new possibilities for movement, experience, self-knowledge, and intimacy. Thinking with cyborgs means thinking specifically and critically about what kinds of people and bodies we consider "normal." Further, thinking with cyborgs challenges us to rethink the social and physical conditions that constrain our ideas of normality, not only in the future, but in the present.

Feminism and Cyborg Theory

Cyborg theory is a diffuse matrix of philosophy, literary theory, history, anthropology, art, fiction, design, feminism, and more. In fact, outside of this book, the authors avoid the term *cyborg theory* because it suggests a field of study that is more unified than it really is. Sometimes the "theory" of cyborg theory explicitly draws on scholarly traditions, which we will refer to throughout this book to situate the cyborg issues we present. At other times one might have a cyborg encounter: more of a feeling or experience of transgressing categories of boundaries when encountering a creative work or tool. For example, we are reminded by our partial machine nature when we put on our eyeglasses in the morning or take a daily medication to calibrate an internal hormonal system. Or a work of art, like Brian Eno's algorithmically programmed 1975 composition *Discreet Music*, might evoke both an emotional, embodied response and urge to figure out the machine logic that produces it.

As feminist scholars, we are both particularly interested in the contributions of feminist thought to this conversation. Feminists' engagement with technology has changed with the times in ways that have, intentionally or unintentionally, mirrored the issues raised by cyborgs and cyborg theory. In the so-called second wave of feminist activism in the 1960s and 1970s, for example, white feminist concerns with child-rearing and women's duty to oversee the domestic space translated into the hope that machines would one day take over gestation and childbirth. In the 1980s, worries about the Cold War and economic imperialism served as a backdrop to Donna Haraway's use of the cyborg as a feminist figure in the canonical 1985 essay "A Manifesto for Cyborgs: Science, Technology, and Socialist Feminism in the 1980s" (also sometimes referred to in short in this text as the "Cyborg Manifesto"). In the 1990s and 2000s, with the institutionalization of "third-wave" feminism, critiques of white feminist thought from nonwhite perspectives offered alternative genealogies for imagining how humans could become—or already had become—cyborg. And in recent years, anxieties about digital surveillance and the possibility of subverting digital systems has reinforced the need for vigilance and care in the embrace of new and promising technologies.

Haraway's "Cyborg Manifesto" is still a key touchstone in cyborg theory.[11] In the essay, Haraway argues that the "figure" of the cyborg offers a way to leave behind thinking that demonizes technology while continuing the feminist struggle for liberation from gender, sexual, racial, colonial, and economic oppression. By honestly recognizing our partial perspectives on the world and relationships with technology, Haraway proposes, it might be possible to build better affiliations that rely less on systems of classification and separation. Cyborgs offered Haraway a way to think about identity and politics without relying on essentializing categories like sex and race. Cyborgs can be real people, she claimed, who exist in the spaces between nationalities, economies, and ethnicities by virtue of their participation in global economic systems that blur the boundaries between nation-states. In addition to hopeful theorizing about the future of high-tech cyborgs, Haraway recognized women like multilingual textile workers in Latin America and circuit board assemblers in Southeast Asia as already cyborgs.

Feminist theorists had a variety of responses to Haraway's idea that feminism and technological change are deeply, and potentially positively, linked. Some responses, such as the essays collected in the 1996 volume *Between Monsters, Goddesses, and Cyborgs: Feminist Confrontations with Science, Medicine and Cyberspace*, sought to work out the philosophical and ethical consequences of the idea that feminists should consider technology as an aid, rather than a challenge, to feminist thought and organizing.[12] That collection was strongly shaped by ecofeminism, a perspective that associates women with nature and sees women as the best spokespeople for environmental issues. Against this background, many of its contributors pondered whether the changes already triggered by technology in the 1990s were entirely desirable, especially those related to childbirth and pregnancy. The anthology illustrates that cyborg antiessentialism and the celebration of new technologies at times caused strain between cyborg feminists and other feminists.

Other responses reframed the engagement between the figure of the cyborg and the writing, work, and lives of women of color feminists. Chela Sandoval, for example, questioned Haraway's use of "third world women" as an instance of cyborg consciousness.[13] She wonders if the celebration of cyborg consciousness of women of color working under exploitative conditions goes too far in making claims about their agency, when in fact they are subject to multiple oppressions. At the same time, she points out even more ways in which so-called third world women already inhabit a cyborg consciousness, suspended between nationalities, indigeneity and the cultures of colonizers, and linguistic fluencies. In this view, women of color and historically colonized women already inhabit cyborg consciousness and might be considered the first cyborgs.

Today, cyborg feminism aims to create a world that is more open to play, freedom, and making nourishing connections across differences. This effort draws on all the conversations sketched out in this introduction as well as many more that you will learn about in the coming chapters of this book. The roots of cyborg thinking will always be in the military science of the Cold War, and technological advances will always be haunted by the possibility that they may be used to sort and selectively oppress different groups of people. But technology need not be opposed to social progress and liberation. Its power can be harnessed *for* these ends too. The cyborg thus remains a figure positioned in what science studies scholar Susan Leigh Star calls a "high tension zone."[14] Suspended between human and machine, past and future, oppression and liberation, the figure of the cyborg offers hope for escape from harmful human pasts while remaining firmly rooted in complicated earthly politics.

Cyborgs Today

There is a reemergence of interest in many of the classics of cyborg theory, from citations of Haraway's work in popular writing to popular media profiles of the 1990s' cyberfeminist art collective VNS Matrix to new exhibits and festivals about Afrofuturism popping up regularly in New York City, where both of us authors live. At the same time, increased ambivalence about whether digital technology will deliver on the promises of boosters is pitting techno-optimists, like transhumanist Ray Kurzweil and tech mogul Jeff Bezos, against humanists and civil rights workers, like sociologist Ruha Benjamin or the Electronic Frontier Foundation. Transhumanists believe that ultimately humans will leave their bodies and merge seamlessly with computers, extending human life spans and ultimately achieving immortality. As scholars of cyborgs in medicine, technology, and design, we welcome the renewed interest in cyborgs and view the increased scrutiny of technology as a conversation where cyborg thinking is crucial.

One of the most important lessons from cyborg theory is that technology does not develop independently from humans. The cyborg, a hybrid of human biology and electro-chemical-mechanical technology in its original Cold War form, is a frank reminder that technology and humans are connected, and even codependent. While Silicon Valley watchers may see automation as posing an either-or threat to people around the globe—for example, either a global economy without automation and living wages, or an economy with automation and mass poverty—cyborg thinking asserts that both automation and human flourishing can coexist. What either-or thinking misses in this case is that technology, like cyborgs, is embedded in histories and contexts in which politics, ethics, kinship, the law, biology, and other technological systems are all simultaneously important for shaping the future of work in an automated economy. Cyborgs are thus a reminder of the power that human societies have to shape, accelerate, or even slow down the development of technologies deemed to be world changing or risky. Technology *is* us.

We also want to intervene in the kinds of cyborgs being produced using technology in the real world. For instance, to go back to Sophia, we are unhappy to see human gender norms transferred into novel technological systems like AI-driven androids. Since she is a human creation, the way Sophia is designed tells us a great deal about how the designers view their fellow humans. It suggests that Sophia's designers expect humans to act in stereotypically gendered ways as well—a disturbing fact to face after fifty or more years of mainstream feminist agitation against restrictive gender norms around the world. As feminist cyborg theorists, we think it is important to hold technologists accountable when they build human biases into technologies.

Finally, we are concerned with offering alternatives to the status quo that go deeper than many popular quick fixes. We want readers to become vigilant about interventions where technology is expected to be a full solution to problems that are social as well as physical, medical, or technological. This is a flavor of what journalist Meredith Broussard calls "technochauvinism": the assumption that technology can fix all of our problems on its own. History and experience prove this assumption to be false.[15] For example, adding a new algorithm to an AI system intended to correct one form of bias among many that are possible might create new forms of bias. For that reason, each chapter includes several illustrations of cyborg thinking and action that we believe will inspire readers in their own lives.

These interventions are vitally important for building a future that is more just and inclusive, yet still playful and innovative. With the rise of "future of" conferences and corporate slogans, an efflorescence of science fiction writing, film, and television from a more diverse cohort of creators than ever before, and a shared feeling that the widespread availability of portable, powerful computers is accelerating social change, the future has never seemed closer. Yet we do not often enough have the time or space to stop and ask whether the future we are creating is the future we want, and whose interests it ultimately serves. The cyborg is a complicated, noninnocent figure that connects history to the future, both mapping possibilities for the future and reminding us where we come from. Thinking like a cyborg forces us to reckon with the oppressions and multiple, partial perspectives that are always part of the human world, including our technological inventions.

Mapping the Conversations

Any project that aims to fully catalog the wide range of scholarship, art, activism, engineering projects, and commentary that draws inspiration from cyborgs is destined to fail. Therefore in this short book, our aim is to offer an intervention into current conversations and debates about cyborgs focused on three topics of current significance: the future of work, disability and reproductive justice, and cyborgs in art, fashion, and design. Rather than supplying a complete genealogy of each of these topics, each chapter will either analyze and situate a core text, like the later chapter on Haraway's "Cyborg Manifesto," or a small number of exemplary projects that demonstrate the potential of cyborg thinking, scholarship, and creativity. You might consider this book an amuse-bouche of the cyborg universe.

The next three chapters examine topics of timely concern along with how a cyborg perspective opens up new solutions and critiques. After that, two chapters delve deeper into scholarly conversations about cyborgs and cyborg theory. Then we wrap up with a statement of principles for cyborg thought in the twenty-first century, a glossary, and resources for further reading.

Chapter 2, "Cyborg Labor," discusses how a cyborg perspective can reshape contemporary debates about technologies such as automation, robots, and AI in terms of influencing the social and economic context around the future of work. These debates frequently center around two dominant paradigms. First, many argue that technology will *inevitably* replace workers, leading to the end of jobs and radical changes in public policy such as universal basic income. The second common vantage point is that as automation eliminates some jobs, humans will find other things to do. This is expected to result in a wide variety of new kinds of jobs, roles, and professions. By contrast, the cyborg can help us recognize that both of these predictions will likely come true, remaining in tension for some time to come. A cyborg perspective can also help illuminate the many assumptions each of these simple narratives often make, including about whose labor is over- or undervalued, and what the timeline might be for broad-scale changes in how and why people work. The real question might be, How can society better address already existing issues of social and economic inequality?

In the third chapter, "Cyborg Bodies," we use the figure of the cyborg to explore the bodily experiences of disability and gestation. From cochlear implants to prosthetic limbs, and from pacemakers to insulin pumps, disabled people have turned to cyborgs in order to make sense of their identities and politics. Disability scholars and activists have playfully appropriated the cyborg to describe some of these identities by coining terms like *cripborgs* and *tryborgs*. As a greater number of people use self-tracking devices, like wrist-worn sleep and fitness trackers, some aspects of the disabled cyborg experience have become more common. In addition to disability, cyborg thinking allows for a new critical perspective on the technological advancements surrounding pregnancy and childbirth. Furthermore, a cyborg viewpoint on human gestation reveals new terrains for political struggle and the liberation of gestators around the world that extend cyborg thinking into reproduction. The cyborg helps to navigate these tricky landscapes of bodily experience.

In chapter 4, "Cyborg Culture," we trace the cyborg's trajectory through cyberfeminist art, Afrofuturism, music, fashion, and design. Channeling the muse of the cyborg—as content, method, or even themselves "becoming cyborg"—many artists and designers have experimented with hybrid images, languages, and materials. For instance, in the fashion world, Comme des Garçons founder Rei Kawakubo's *Art of the In-Between* show at the Metropolitan Museum of Art included 140 examples of her work since the 1980s, organized around breaking down nine dualistic themes that echo Haraway's critique of dualisms in "A Manifesto for Cyborgs." We also turn to electronic music and the Afrofuturist tradition, which while it predates feminist conversations about cyborgs, has made many contributions to the art and design of contemporary cyborgs in keeping with the themes we develop throughout this book. Afrofuturist films like the blockbuster *Black*

Panther, for example, are rooted in an Afrofuturist understanding of high-tech futures that coexist with diverse cultures and traditions rather than erasing them in the name of progress. Korean artist Lee Bul creates sculptures, installations, and multimedia projects about monsters, cyborgs, and transhumanism. This chapter engages with art, music, and design as modes of inquiry that have the potential to generate rich and diverse theories, concepts, and practices about the cyborg.

We devote the fifth chapter to a discussion of Haraway's "Cyborg Manifesto." This essay is the launching point for many feminist conversations about cyborgs. While important, it is a difficult and complicated text. Because Haraway's piece is so ubiquitous in feminist cyborg theory and practice, we decided that it is valuable to unpack this essay at length. Haraway's manifesto sets out to reclaim the cyborg from its military institutional history and provide a guide for using the cyborg as part of a feminist technology research agenda. The critiques that this work sets forth against unthinking technological progress, the speed and expansion of high-tech labor, and retrenchment of gender roles during the Reagan years of the 1980s all inform our own analyses of labor, cultural production, and bodies throughout this book. Ultimately, the cyborg politics that Haraway lays out in this essay are also the politics of what it means to be human in a rapidly changing, technology-infused world.

In chapter 6, "Troubling Cyborgs," we look at the points of disagreement around cyborgs within scholarly conversations. One trademark of the cyborg as a cosmopolitan creature is that it has traveled far and wide in scholarly discourses as well as in the public imagination. As the concept has moved beyond feminist science studies, scholars have raised new concerns that must be addressed for the cyborg to continue to be relevant in the twenty-first century. For instance, throughout history and still continuing today, women, people of color, people with disabilities, LGBTQ people, and many others have been systematically dehumanized through linguistic, legal, and socioeconomic structural inequalities. While our understanding of the cyborg centers the critical and political potential of the concept, at the same time it unavoidably develops out of white, Western feminist intellectual traditions. In this chapter, we grapple with the critiques and update the figure of the cyborg to meet the challenges of the contemporary moment. We seek to create a more inclusive framework for understanding the interdependencies of humans and technology in the years to come.

In the conclusion, "Cyborg Futures," we lay out ten principles for cyborg theory in the twenty-first century. We take our cues from all the sources we discuss throughout the previous chapters of the book: scholarship in the humanities and social sciences as well as art, design, film, and real-world struggles around labor, disability, and reproductive rights. These principles are not exhaustive but instead suggestive of what a new politics for living well with technology could—or perhaps should—look like. If we are to design future systems that adequately address inequality, cyborgs are an essential guide.

We hope this book will be a useful provocation for thinking differently about technology and ourselves. Further, we trust it will help jolt our readers out of either-or patterns of thought, decenter Western, ableist, imperialist, white supremacist, single-mindedly capitalist, and patriarchal and misogynist views of the urgent problems to be solved, and critically reflect on how technology and humans are jointly implicated in creating the conditions for justice and multispecies flourishing. We hope that readers will learn about new conversations and literatures, and encounter new ways to approach, analyze, and solve the problems they encounter in their work. By rerooting the figure of the cyborg in feminist traditions and updating the "feminism" in which it is based, this book has the radical ambition of helping readers find some footing for building modest utopias.

CYBORG LABOR

"We are not robots," shout the striking Amazon warehouse workers. Many work over sixty hours per week in dangerous conditions on little sleep. They hold signs with a recognizable sad face, fashioned by turning the characteristic brown cardboard boxes with the company's logo upside down and adding two black dots for eyes. In 2022, over one million people worked for Amazon in the United States, making it an important case study in understanding the changing relationship between people, technology, and the social and economic conditions that shape work and the workplace.

As humans, we will spend over one-third of our adult lives, or over ninety thousand hours, working. This chapter develops critical cyborg literacy in order to understand the changing conditions around work in neoliberal capitalism, whether the labor takes place in offices, factories, warehouses, farms, or other everyday settings. Critical cyborg literacy enables society to make more responsible decisions about when, where, and how to integrate technologies such as automation, robots, and AI into our work and workplaces. It also guides decisions about when to actively resist such applications.

There are several common narratives about automation, robots, AI, and work that appear frequently in the mainstream media as well as in academic and corporate discourses. First, there is the worry that robots will replace humans, emphasizing competition along with the elimination of many jobs, roles, and professions. Second, there is the hope that robots will cooperate with humans, stressing collaboration and the expansion of many new kinds of jobs, roles, and professions. And third, there is the expectation that robots will reconfigure existing jobs and create new ones. It is easy to become convinced of any of these narratives based on only a few examples. In this chapter, we examine multiple scenarios side by side to understand the broader pressures, incentives, and desires behind these stories about the future of work.

In order to balance our discussion of the broader trends and specific instances, in this chapter, we center the role that Amazon has played in shaping everyday life over the past twenty-five years. For some, Amazon has become an essential service for the speedy delivery of household goods and in-home assistance through the company's

Alexa device as well as audiobooks, music, and even entertainment. The company has also expanded to groceries and brick-and-mortar storefronts through its acquisition of retailer Whole Foods. With Alexa waking us up each day, our meals provided through Amazon Prime grocery delivery, and the latest Amazon Originals series like *The Lord of the Rings: The Rings of Power* (2022) putting us to sleep at night, perhaps it is not an understatement to say that we literally eat, sleep, and breath Amazon.

For others—namely warehouse workers and delivery drivers—Amazon is an exploitative employer. During the COVID-19 pandemic, the company's profits soared after it captured at least 50 percent of the total online retail sales while at the same time its warehouse workers were striking. Amazon has significantly impacted the job market by displacing retail workers, and in exchange, created new jobs but with much worse working conditions.

The first question that a cyborg perspective might ask in the face of any claim about the potential benefits and/or harms of technology is "For whom?" Without more specific information about who will be impacted by any decision about technology, it is impossible to draw any conclusions. So we must ask, Who will be impacted? Who has the power to make decisions? Who can refuse?

Of course, this distinction at the heart of the "for whom" question is not binary or black and white. There are likely people who are both consumers of Amazon products and services as well as workers for one of its many divisions. But the people who are likely to be in either of the above groups are likely to map onto existing global inequalities—wealthier people are more likely to enjoy the benefits while poorer people are more likely to suffer the consequences.

Amazon's effects extend beyond its direct customers and workers. These are some of the broader trends that we address in this chapter. Amazon plays an important role in shaping global supply chains, logistics, and distribution networks. These networks influence the timing and availability of goods beyond any single company. Amazon has also pioneered the Mechanical Turk platform, which allows employers to break up big jobs into simple, repetitive microwork tasks. This style of working has been recognized as exploitative for at least two hundred years, and with digital technology, now has a much wider reach. Finally, Amazon is at the center of debates around the use of robots in its warehouses. Some worry that Amazon's example will lead to further layoffs among warehouse and logistics workers, or wonder whether it will pave the way for more collaborative work with machines over the long term. Some even expect that Amazon will likely be an early adopter of biotechnology enhancements to give workers seamless access to work sites and increased efficiency.

For all of these reasons, Amazon is an excellent site for the discussion of the future of work. The Amazon case allows us, as the writers of this book, to develop and apply critical cyborg literacy. Critical cyborg literacy offers necessary critiques and potential pathways for resisting the most harmful applications of technology. How can we instead imagine technologies that support liberation and justice?

In order to illustrate this, we integrate examples from several films—*Sleep Dealer*, *The Zero Theorem*, and *Sorry to Bother You*—that help to imaginatively critique existing narratives, presenting cautionary possible scenarios and bringing these narratives to life in a more experiential sense in order to more deeply engage with these futures of work. Together, these three films foreshadow a range of existing and emergent individual experiences, collective work arrangements, and organizational forms that are supported through the use of digital technology in concert with the power relations structured by the current socioeconomic conditions. These films are useful for identifying both the implications of dominant narratives around the future of work and counternarratives. In addition, each film offers stories that explicitly center issues of gender, race, and immigration status as one aspect of defining the experience of work.

By analyzing a real-world corporation alongside recent films, we aim for a playful engagement that draws on the strengths of storytelling and narratives—real and imagined, factual and fictional, rational and embodied, plausible and fantastic—for critique and resistance as well as imagining alternative possible futures. It is in conversation with the latest forms of scholarly practice, art and design methodologies, activism, and public engagement too. Critical cyborg literacy offers a useful starting point that helps make sense of (and complicate) these narratives. Moreover, it provides ballast for living with dissonance, contradictions, and the gaps between broad claims that often dominate mainstream media headlines and the lived realities that people, especially those in oppressed and vulnerable communities, must contend with every day.

The cyborg is an essential concept for understanding automation and the future of work because it urges us to question boundaries and borders, like those between human and machine, domestic and foreign, employee and freelancer, employee and customer, and work and life. In this chapter, we explore a range of questions about the nature of cyborgs at work. Where do cyborgs work? What do they do? Who compensates them for their labor? How are they paid? How do they pass the time? When does their day begin and end? Who are their colleagues? Who are their bosses?

This chapter analyzes dominant discourses along with less well-known issues and alternatives around automation—from robots to AI—and the future of work. These shifting boundaries and borders have implications for how much workers are paid, whether they are considered

By analyzing a real-world corporation alongside recent films, we aim for a playful engagement that draws on the strengths of storytelling and narratives for critique and resistance as well as imagining alternative possible futures.

The cyborg is an essential concept for understanding automation and the future of work because it urges us to question boundaries and borders.

employees, and even whether work is compensated at all. Critical cyborg literacy is essential for holding these multiple narratives in mind while asking political questions around whose labor is valued or undervalued, what temporal frames matter, and how society might address broader issues of social and economic inequality.

Automation Takes Work

The Museum of Modern Art's *Good Design* exhibit in 2019 featured the "Quik-Suds Semi-Automatic Dishwasher," designed by Harold Osrow in 1948. The device consisted of a hose that could be attached to a kitchen sink faucet and was made of plastic, rubber, nylon, and steel. It was a winner of the museum's Good Design award when it was introduced. The device's box advertised that it "Makes Dishwashing a Pleasure" because it solved the "problem of dishpan hands" by keeping hands dry and using a button to get "instant" suds and rinse dishes.

This example is interesting because it illustrates the ways in which technology, gender, and domestic labor were understood at the time. Specifically, by advertising a "semiautomatic" technology, the device promises to make the lives of women confined at home in the post–World War II period of the late 1940s and early 1950s more enjoyable. Many women became employed outside the home during the war by necessity, but then returned home after the war as full-time housewives.

So for whom are "dishpan hands" considered to be a problem and why, if at all? Upper-class women were likely to have servants helping them at home so this device would more likely benefit women who took on their own housework. This new device might have even justified a shifting of responsibilities in which women ultimately took on even more tasks inside the home as those tasks became more automated, as Ruth Schwartz Cowan's *More Work for Mother* has argued.[1] Today, in-home assistants such as Amazon Alexa are playing a similar role in increasing the burden of domestic labor on women, as Yolande Strengers and Jenny Kennedy described in their book *The Smart Wife*.[2]

While these instances of automation tether women's work to the home, Liz Faber writes in her book *The Computer's Voice* that "in the late 1980s and early 1990s, computers were conceived in advertising as women's assistants in a 'postfeminist fantasy,' allowing them to work outside the home while also maintaining the traditional domestic roles of housewife and mother."[3] Critical cyborg literacy demands that we ask such questions about technology, gender, and labor, which we will return to in the next chapter, "Cyborg Bodies."

While low tech compared to today's digital technologies such as AI and robots, the semiautomatic dishwasher example illustrates a number of important themes that we look at later in this chapter. First, automation still requires human labor, but the labor and laborer may be different than in the past. This is a case of what sociologist of science Bruno Latour has referred to as the "missing masses" in which jobs once done by humans are displaced by technologies.[4] In addition, gender, race, and class play a significant role in determining the jobs that are displaced, the jobs that are created, and, ultimately, who might benefit from these changes. Second, automation is often equated with magic that occurs instantly at the press of a button (rather than the reality that humans will still be doing much of the work).

Narratives about automation and the replacement of human labor have a long history. One useful starting point is the Luddite revolt in the early nineteenth century. Calling someone a Luddite typically means they are opposed to new technology or progress. The term, however, refers to a real group of textile workers in the nineteenth century United Kingdom, led by General Ned Ludd, who destroyed machines in textile factories because they believed that automation would replace their jobs.[5] Between 1811 and 1817, a group known as the Luddites staged riots to break the machines in the textile industry, which they blamed for wage reduction and unemployment. Thus the Luddites were not inherently antitechnology or antiprogress but rather opposed to the power of employers and government officials who had deprived them of their ability to make a living.

Many of the power dynamics that are embedded in the use of automation today recall early twentieth-century Taylorism too. This approach to management was named after Frederick Winslow Taylor and prized the scientific study of work processes to increase efficiency.[6] Each task would be studied and broken down, and each part of the task was optimized so that workers could work more efficiently. Machine systems also enforced the increasingly rapid pace of the simplified, perfected work that human laborers did. Today, digital surveillance, microwork, just-in-time scheduling, so-called algorithmic bosses, and other uses of technology in combination with the current form of extreme neoliberal capitalism allows for the emergence of a kind of new Taylorism.[7]

Contemporary headlines about the "rise of robots" and future of work proclaim "the end of jobs as we know them," implying that robots will eventually replace all forms of human work.[8] These stories assert that technology is destroying jobs and provide reasons why we should fear the robots that will take our jobs. Sometimes they depict robots as "immigrants from the future" and describe new kinds of automated work to which "humans need not apply." These headlines are technodeterministic because they suggest that technology will override human agency to drive progress without our oversight. They are problematic because they displace power, responsibility, and agency onto machines. Instead, critical cyborg literacy acknowledges that robots are embedded in the socioeconomic systems created by humans, specifically a particular form of late stage neoliberal capitalism.

According to the utopian version of this narrative, humans will eventually be completely free from work, enjoying lives of leisure. A darker, more dystopian version is that professional, middle-income, and low-wage work will be automated and replaced, but without adequate social support, there will be class warfare. This has prompted discussion of policy approaches such as a universal basic income in which all citizens would receive a small amount of money to offset their living expenses. Yet as many scholars have pointed out in recent years, in many cases when humans are replaced by "automation," someone still needs to do the work. And most automation is not a glamorous process of interacting with an attractive cisgender robot with a pleasant female voice but typically much more mundane everyday situations.

Let's take a simple example: automated teller machines (ATMs) or self-checkout at the grocery store. In both cases, new technologies were used to replace and reconfigure human labor. Before ATMs, it was not possible to deposit or withdraw money from the bank without going to a physical branch and meeting with a human teller. But now, customers take on the work of depositing or withdrawing the funds for free, as unpaid labor. A machine has been introduced, but there is work to be done in order to use the system. Similarly, at the grocery store, the customer can now do the work of cashiers and baggers—again without pay. Occasionally a manager is required to approve certain purchases or troubleshoot the system, but overall, far fewer human workers are required. And in online purchases, we have become accustomed to doing all the work of checking out and paying ourselves.

In these instances, while human labor is still required, this labor has been off-loaded onto the customer, who willingly does it for free in the name of convenience, speed, and efficiency. The labor has shifted across the boundary between employee and customer. In the process, the company saves money and probably reduces its employee head count.

This is not unlike the work that we perform every day for free when we use search engines, maps, and applications that use our data to optimize software performance and corporate operations. In these cases, claims about the magic of automation obscure the fact that humans are doing a great deal of work that helps to keep digital systems and software running. Human work to identify and classify information is hidden within the normal operation of Google searches and Facebook's social media content moderation.[9] Here, while technology companies claim that their processes rely on their prowess in AI and machine learning, they are often hiring companies to outsource tedious, boring, and even traumatizing labor to contract workers around the world. For example, in order to rank websites, Google uses third-party human raters at third-party companies to evaluate the quality of search results. These workers do not have the same status or benefits as full-time Google employees.

Next we turn to a discussion of distributed work, algorithmic management, and biotech enhancement as portrayed in three recent films to bring to life these dynamics around the transformation of work.

Cyborgs at Distance: Remote and Distributed Work

The 2008 film *Sleep Dealer* directed by Alex Rivera tells the story of Memo Cruz, a Mexican construction worker based in Tijuana near the US-Mexican border in the near future. Since migrants cannot cross the border in order to work in the United States, Memo works in a "digital factory" where he connects his body to a robot in order to build apartments remotely. *Sleep Dealer* represents the remote and distributed real-time digital control and management of people, things, and machines in global logistics and supply chain networks.[10] While the film is over ten years old, it still depicts a relevant scenario. The narrative portrays a dystopian version of the future of work in which work is remotely controlled and globally distributed. In the movie, this arrangement of labor satisfies corporate needs to obtain a limitless amount of low-cost labor as well as government desires to control and prevent immigration.

In such a future, the current debates about the future of work that we are having at present (including those around border walls and Brexit) would be moot. In this scenario, even construction work (arguably one of the more materially embodied forms due to the current requirement of physical bodies as well as physical materials such as wood, cement, and steel) is reconfigured politically, economically, and materially through automation. Physical labor (such as farmwork, maintenance work, and janitorial work) and care work (such as nursing and teaching) are often described as forms of labor that are less likely and/or even impossible to replace through automation. Yet there are still many instances of technologies that are being prototyped in order to do just this. Think, say, of robot vacuum cleaners, food delivery vehicles, and robotic in-home care attendants that are frequently showcased as techno-optimistic examples for the future.

Furthermore, with the emergence of virtual, remote, and distributed work made possible by portable computers and broadband networks, employers can hire a greater number of on-demand, contract, temporary, and other types of independent workers. This allows employers to reduce the cost of labor by dynamically adjusting it more flexibly with respect to other costs. It also allows employers to off-load the physical costs of maintaining their fixed office spaces and other associated connectivity expenses such as computing equipment and cell phones to employees. Instead, some of these costs are paid by employees, some of which work from home or in flexible shared offices and coworking spaces with shorter-term leases and rental agreements. In the current economy, which is designed to favor fast-paced change and disruption, long-term fixed expenses such as those for labor and real estate are converted to flexible ones.

An alternative set of headlines frames the issue a bit differently. In these visions of the future of work, humans and machines work collaboratively. Machines replace some kinds of work and workers but, overall, work is reconfigured so that humans can focus on the creative knowledge work that robots are not capable of doing (yet). For example, a recent set of headlines identifies AI as helping humans invent new flavors of beer and scents of perfume.

On September 10, 2017, the *New York Times* showcased a video titled *Robots and Humans Team Up at Amazon* that featured footage from Amazon's facilities in New Jersey, lauding the collaboration and cooperation between warehouse workers and robots. At the time, the company had 125,000 human warehouse workers in the United States and 100,000 robots. In the video, an Amazon associate, Michael Tyre Sr., explains that robot palletizing—using the robot to put the material on the pallet—is more efficient, allowing humans to do more things: "It allows you to be a team member with the robot, you're a team member with your colleagues. . . . It's a team effort between the humans and the robot. We all work as one."[11]

This video casts a positive light on automation, assuaging our fears about the potential threats and risks to jobs, and even going as far as suggesting that humans and robots can be thought of—by employers as well as employees—as team members.[12] This framing serves Amazon well because it minimizes potential negative aspects of automation in warehouse work. The focus on speed and efficiency that is possible with increased automation also means that Amazon warehouse workers have been injured more often at warehouses with robots, and these injuries occurred at twice the rate than at other companies.[13]

While dangerous working conditions created by forcing humans to work like machines are clearly not acceptable, perhaps it is not so unusual that people consider robots as team members. Think, say, of the other technologies that we use every day in the home such as iRobot's Roomba vacuum or Amazon's Alexa, which science studies researcher Laura Voss has studied in her book *More Than Machines*?[14] Voss shows that laypeople as well as expert roboticists themselves attribute forms of agency and animacy to machines, like referring to them by gendered pronouns and giving them names.

Moving on from distributed work and human-robot collaboration, the next example in this chapter is about the ways in which work is managed through software interfaces creating the perception of algorithmic bosses.

Cyborgs on Alert: Alienated and Underpaid

Terry Gilliam's 2013 satire *The Zero Theorem* offers another vision of a future of work in which a computer programmer named Qohen Leth is remotely managed by an algorithmic boss. While ostensibly about the meaning of life, the film supplies a glimpse into the isolation and alienation of the digital piecework that makes up the global economy. In the film, Leth, a white knowledge worker, copes with persistent digital surveillance, alerts, and warnings while experiencing existential dread that life has no meaning.

These alerts are exaggerated in the film, but in some ways they illustrate the psychological impact of "algorithmic bosses" on gig workers such as Uber drivers, Amazon Mechanical Turk workers (also known as Turkers), and Deliveroo and GrubHub drivers. It is not difficult to understand the oppressive nature of digital systems as they are being introduced in a wide range of professions in order to increase compliance with rules and regulations (and decrease human error). Under the guise of "digital transformation," many professions are adopting digital systems in order to manage everything from digital manufacturing that relies on virtual and augmented reality to using sensors to track the physical locations, bodily movements, and productivity of employees as well as the status of goods as they move throughout global supply chains. There are many studies within the field of human-computer interaction that demonstrate the intrusiveness and negative impacts of technologies such as ankle bracelets (worn by formerly incarcerated people), medical devices, and even in-car systems for preventing drunk driving (which have been shown to cause accidents).

In these examples, workers are managed by algorithms that assign tasks based on criteria that are black boxed or unknown to the workers themselves. While sometimes described as freedom to set one's own hours and work without a boss, a series of recent books such as Mary L. Gray and Siddarth Suri's *Ghost Work*, Alex Rosenblat's *Uberland*, and Brooke Erin Duffy's *(Not) Getting Paid to Do What You Love* have described and examined the ways in which the gig economy is restructuring work to the detriment of workers.[15]

While one might hope that assigning rote work to machines would allow humans to take on more stimulating and creative work, this is not the case. Rather, with the expansion of automation, robots, and AI in modern workplaces, humans are required to act more like machines. Robots in Amazon warehouses and the Amazon Mechanical Turk reduce human autonomy and decision-making. Workers are instead more closely and frequently alerted, disciplined, and surveilled by machines.

According to the *New York Times*, human workers have "Amazon's software looking over their shoulder." In an Amazon warehouse, where job roles range from pickers and packers to water spiders and stowers, the stowers have the most autonomy and decision-making ability. "If a stower attempts to stick, say, a bottle of sunscreen in a bin next to one that already has such a bottle of sunscreen in it, the bin will light up to signal that this is not allowed. Adjacent bins with identical items can confuse the pickers."[16] Overall, in contrast to narratives about increasing time for human creativity, robots have increased productivity, but also made jobs more repetitive for humans. These narratives are somewhat less technodeterministic in that they emphasize the coevolution of humans and machines (as opposed to the revolutionary replacement of humans with machines). At the same time, the ideologies behind these narratives, such as how technology is linked to modernity and progress, and what the envisioned role of humans is—which are often embraced by technology companies, computer scientists, and scholars alike—mask the complicated politics.

"Smart" Cyborgs: Biotech Enhancement of Workers

So far in this chapter, we have discussed the reconfiguration and distribution of work globally, use of algorithms and digital platforms to manage and control work, and collaboration with robots on the floor of factories and warehouses. We have illustrated how cyborg theory might make sense of these dynamics. Going beyond these themes, another area of significance is the biotechnological enhancement of workers.

In recent years, there have been examples of companies that are actually turning their own employees into cyborgs by embedding them with radio-frequency identification chips. For instance, according to the *Atlantic*, as of 2017 several companies had already deployed this technology, which purports to make building access and payment easier (but could be used for surveillance too).[17] While not mandatory today, in the future, without political resistance, it is possible that some workers will be required to have certain enhancements or technologically augmented capabilities.

Such use of technology to enhance the bodies and minds of employees is foreshadowed in *Sorry to Bother You*. In this film, a 2018 dark comedy by director Boots Riley, the character Cassius "Cash" Green is a Black telemarketer for a company called RegalView in Oakland, California, where he is living in his uncle's garage with his girlfriend, Detroit. By using his "white voice," he is promoted to a Power Caller, which allows him to afford a new car and apartment. Yet at a party with the CEO of WorryFree, one of RegalView's clients, he learns that the company plans to create strong, obedient half-horse, half-human "equisapien" workers. Cash is offered $100 million to lead the transformation by becoming an equisapien.

Sorry to Bother You also engages with discourses around slavery and transhumanism. It suggests that digital and biotechnology enhancements will be used to create augmented humans that can be superworkers for the benefit of corporations. These superworkers will not have control over their own lives and bodies, drawing parallels to the historical enslavement of people of African descent in the United States. Since Cash ultimately betrays his friends by crossing the picket line on his way to a more prestigious position with RegalView, the film also emphasizes the extreme individualism of the current economy in which solidarity with workers and unions have weakened in recent years. Labor rights could serve as a bulwark against abusive applications of transhumanist ambitions, it suggests, but only if workers intentionally choose solidarity and unionization.

Films such as *Sorry to Bother You* use extreme scenarios to provoke critical thinking about current situations as well as the kinds of futures that we are ultimately willing to accept. The next section suggests that both technologies and socioeconomic conditions can be transformed through efforts to build worker solidarity, such as the recent unionization campaigns at Amazon, Starbucks, and other large employers like universities.

Reimagining the Future of Work

In response to the challenges introduced by automation, robots, and AI, there is continued work by the social and economic justice communities to build worker solidarity and to organize workers to strike against exploitative employers, including fast-food chains, retailers, and even technology companies themselves.

One early effort was the Fight for $15 campaign for a $15 minimum wage that has gained traction in many states over the past ten years.[18] While Amazon announced a $15 minimum wage in fall 2018, at the same time it has been sharply criticized for cutting back on hours and the declining work conditions at its companies, including high-end grocery and retail chain Whole Foods.[19]

Over the past few years, organizers have led protests in order to disrupt companies including Google, Facebook, Microsoft, Amazon, and Uber. In addition to wage inequality and work conditions, these protests have emerged in response to sexual harassment and the #MeToo movement, support of drone warfare, and selling services to police forces and defense agencies. Yet these strikes have not been limited to technology companies. Other businesses such as hotels and real estate companies have also faced protests by employees and residents that disagree with plans to use technology to replace jobs and track people.[20]

Furthermore, there are cases of technologies that have been designed specifically to support fairer working conditions. One example is Turkopticon and a related project, Dynamo, which serve as communication platforms for discussions among the Amazon Mechanical Turk community.[21]

Because workers are physically isolated, and their work is fragmented into a small task or set of small tasks, they have little ability to build relationships with other workers. These platforms allow Turkers to share information about wages, exploitative employers, and undesirable tasks.

Another illustration is the modification of just-in-time scheduling, which uses algorithms to manage the schedules of workers. Just-in-time scheduling makes it difficult for workers to predict their schedules including their needs for transportation, childcare, and other aspects of their lives. So retail workers might end a shift at midnight only to find out that they have been scheduled to return at 5 a.m.—a situation known as "clopening" (referring to closing and opening). In response, a company called Kronos has created a fairer just-in-time scheduling software.

Over the past few years, relationships between social and economic justice advocates and technology for social good leaders have deepened, which has spurred ventures such as the Workers Lab to fund innovative uses of technology to protect, organize, and advocate for better conditions for workers. Another important community has formed around platform cooperativism, which aims to support the creation of cooperatively owned digital platforms and organizations such as FairBnb rather than their privately owned counterparts.[22]

Both attempts to resist exploitation as well as redesign specific technologies are essential. But they face an uphill battle against the shape-shifting nature of contemporary capitalism. Capitalist prioritization of accumulation, speed, and efficiency pervade more and more institutions, and largely determine the rules of the broader socioeconomic system. We hope a critical cyborg literacy perspective on technology-aided work can help seed new ways of thinking and being that allow us to reimagine the relationship between humans, technology, and the economy writ large.

Rather than focusing on control and surveillance, for instance, our dominant narratives about automation, robots, and AI could be reframed around imagining alternative economic forms with an emphasis on cooperative models for sharing power, equitable pay structures, and more just decision-making.

Conclusion

This chapter has engaged with narratives around automation, robots, AI, and the future of work, arguing that critical cyborg literacy is a useful approach to redefine, resist, redesign, and reimagine the ways in which humans, technologies, and the larger socioeconomic system might create modes of working and living in the future. We began with three scenarios about the future of work—one framed around the replacement of human labor, one around collaboration, and another that suggests a reconfiguration in which both of these scenarios is partly true. By using evocative films as well as empirical examples, we analyzed these specific narratives and scenarios in the context of a large corporation such as Amazon, and illustrated some of the nuances present within these narratives about automation and the future of work.

Different workers experience different kinds of work differently. Often, these differences are the product of gender, race, class, ability, sexuality, immigration status, and so on. To put it more clearly, the most vulnerable and precarious workers will continue to be exploited—and may even be further exploited—as society adopts automation given the existing socioeconomic conditions under neoliberal capitalist structures. Furthermore, the contexts, places, and timescales in which different people might experience change, disruption, stagnation, and transformation are diverse too. It is also possible that there are many alternative potential futures of work that have not yet been imagined or invented.

This chapter has deliberately left out some important discussions around labor—specifically, surrogacy and reproductive labor—that are core to a feminist approach to thinking about technology. We will turn to this topic in the next chapter.

CYBORG BODIES

Imagine that it's the year 2374. Far from Earth, a spaceship crew chances on a ship full of half-organic, half-machine beings bent on assimilating every starship, planet, and species they encounter. For the moment, though, you're allied with them against another species that threatens both your ship and theirs. Their representative, a blond entity who was born human and still resembles a human woman, considers themselves a component of a nine-part unit. They have instant access to the collective experiences of billions of former individuals of dozens of species. Implants on their limbs and face as well as in their bloodstream improve their strength, vision, and healing capacity compared to an ordinary human. Talking to this unit, who is beautiful, alien, and haughty, is like interacting with an upgraded human. The technologies in their bodies would improve your crew's chances of survival and success on its journey, if you could adopt some of it and still maintain your ship's independence.

Now imagine that it's 2 a.m. in the year 2016. A machine connected to your body has been beeping every hour with false alarms about your blood sugar levels. You haven't slept well in months because this *just keeps happening*, no matter how many times you try to calibrate the machine. You're tired and sometimes irritable. Your job is demanding and difficult to do without enough sleep. You like to exercise to relieve stress, but some days you're just too tired to put on your running shoes after being up all night. It's a mundane problem that your doctors have no solution for, but it is significantly detrimental to your life, every single day. While the machine might represent the most current biomedical technology, it feels like you've taken a step backward from other ways you could monitor and manage your blood sugar.

The first story might be familiar to you. It is the first encounter between the starship *Voyager* crew and Seven of Nine, a member of the Borg collective who later becomes part of the ship's crew in the 1990s' television series *Star Trek: Voyager*. Seven of Nine is the quintessential science fiction cyborg. Her body blends organic, human tissue with digital sensory enhancements, shiny metallic implants on her face and hands, and invisible nanorobots that accelerate her body's healing when she is injured. She does not exist, except in a fictional projection of a future galaxy. The actor who plays her, Jeri Ryan, is also slim, white, blond, and North American. Seven's supposed "human side" comes out in later episodes in heterosexual flirtations with other crew members.

The second story comes from Laura's 2017 paper, "Data Rituals in Intimate Infrastructures: Crip Time and the Disabled Cyborg Body as an Epistemic Site of Feminist Science."[1] Laura's account of living as a disabled cyborg takes place in the here and now. The cyborg technology she uses is intended to fit in with other more common technologies such as cell phones, though it is essential to her body's basic functioning.[2] Her experience of living with cyborg technology isn't exciting or sexy. Instead, it is characterized by biological necessity along with a struggle to be heard by doctors and the company that manufactures the device when it malfunctions over and over again. Yet living as a cyborg is also a source of creative potential for Laura: finding imaginative ways to adjust the device, channeling her experience with it into her scholarly research interests, and even collaborating with artists to turn the device's data into a sensory experience of cyborg living that others can partake in.

While popular culture offers up images of cyborg bodies that exceed the capabilities of the human and subsume messy culture to technological solutionism—cyborgs that recall science fiction characters like Seven of Nine—the real-world experience of living as a cyborg is much more complex. In this chapter, we discuss two types of cyborg bodily being: disability and gestation. Both modes of living raise questions about how to live with technology as well as how race, class, gender, nationality, and disability intersect, and how this affects the role technology plays in everyday life. We foreground cyborg bodies in the material world rather than imagined ones in fiction to demonstrate how technological interventions for disability and gestation can both constrain individuals' lives and open up new creative possibilities. Examining such "real" cyborgs offers a more immediate way to appreciate the beneficial effects of cyborg technologies on individuals and society as well as understand the harms they produce in actual people's lives.

We chose disability and gestation for our examples because these modes of being are central to feminist knowledge practices and politics, and we are both feminist scholars. Bodies in general are central to feminist politics, in fact. Feminists have often taken the painful, hidden, and shameful experiences of bodies coded feminine or female as evidence of women's oppression. For instance, sharing experiences about living as a woman through childbearing and menstruation was a way to raise consciousness in 1970s' feminist activism.[3] Armed with collectively shared knowledge about their bodies, women in these groups could then take action through advocacy and the rejection of medical authority; they could also take science and medicine into their own hands.

The more recent intersectional feminist paradigm expands feminist consciousness to be more inclusive of different kinds of bodies and more fully recognize the experiences of women who are in the crosshairs of intersecting forms of oppression—not only gender oppression, but racial, ethnic, national, class, caste, and other forms of oppression. Within this frame, disability and gestation have drawn recent, sustained attention from feminist thinkers because they exemplify how technology shapes the experience and politics of living with/as a body in many contemporary settings where technologies are used to intervene in bodies.

Here we use these two domains of experience to explore how technological interventions on the body can unsettle what we take for granted as the "normal" body. This is a pivotal aspect of critical cyborg literacy. We consider the sometimes overlooked role that technology plays in sustaining bodies and bodily processes that we might otherwise think of as "natural." These issues are not abstract; too often, it is only people with a body considered normal or natural who are awarded full rights to education, economic stability, freedom from violence, and citizenship.

Viewing the body with critical cyborg literacy reveals that living intimately with technology does not automatically mean a person is dependent on it for all things, as able-bodied people sometimes assume about disabled people who use assistive technologies. In cyborg living, technology provides both constraints and opportunities for new forms of autonomy and creativity. As feminists interested in improving how people live with technology, we are fascinated by such creative potential. As you will find in the examples to come, seeing technology as cyborg shapes how one might approach interpersonal interactions, shifts the politics and priorities of technological innovation in the United States, and supplies an anchor for transnational collective labor organizing.

This chapter offers three illustrations of cyborg bodies. First, we explore the idea of intimate infrastructures, which is a useful frame for understanding cyborgs because it offers a way to think about interdependencies between biological and machine body parts without prioritizing one or the other. The machine parts only work when they are part of a body, and the bodies they interface with are sustained by their human-made components. Then we discuss the concepts of cripborgs and tryborgs. These playful adaptations of the cyborg have been invented by disabled people to highlight the identity work that real-life cyborgs undertake as they navigate the world with technology. Finally, we talk briefly about gestational surrogacy, which offers a particularly clear and politicized challenge to the archetype of the natural, technology-free human body. We end by reflecting on the implications of a cyborg perspective for real-world health justice.

In cyborg living, technology provides both constraints and opportunities for new forms of autonomy and creativity.

Intimate Infrastructures

Living as a disabled cyborg means living in close contact
with technologies, from mobility aids to endocrine system regulating devices. For Laura, as a "disabled cyborg,"
both her body and the technology can be understood to
be disabled. Importantly, this understanding resists the
common ideas around the assumption of technological
perfection. Laura calls these cyborg relationships with
technology "intimate infrastructures" in her autoethnographic reflections on managing her diabetes with an
automated insulin pump. She explains that "as a disabled
cyborg . . . my body is networked and dependent on a system of technologies that is fragile, vulnerable, and prone
to break down." This intimacy has concrete boundaries
based on the technological capacities of the devices that
make up her insulin monitoring and delivery system at
any given time. For example, she writes that in 2017, "I
cannot go more than an hour without being plugged in. I
can be no more than twenty feet from the CGM [continu-
ous glucose monitoring] receiver."[4] Moving too far from
the CGM, losing power, or running out of insulin in the
pump prompts the system to deploy a raft of irritating
beeps and alarms, day or night, interrupting sleep, meals,
and professional engagements.

For Laura, then, disabled cyborg intimacy with technology is fraught. The technological infrastructure in-

cludes a "system of technologies . . . the pump, a CGM, a glucose meter, test strips, and software [that runs on her iPhone]."[5] With the system that Laura used from 2013 to 2017, her body and the multiple technological components are interdependent to create a fully functioning system. Her body benefits from the sensor's alarms and insulin delivery because they help her appropriately regulate blood glucose levels. The technological components are dependent on Laura for inputs like drops of blood, charged batteries, and full insulin cartridges. Without the CGM technology, Laura's blood sugar could fall or rise to dangerous levels without her knowing. Without Laura's active management, the system would cease to operate. This interdependence is a key characteristic of the intimate infrastructures that disabled cyborgs live with every day.

But in the newer "smart" system that Laura began using in 2018, its minor (and even sometimes humorous) inconveniences became serious obstacles to surviving everyday life. Her more recent writing illustrates the extent to which the system as assembled is glitchy and harmful. Its various components vibrated day and night in regular patterns that disrupted her sleep for days, then weeks, and then years.[6] Sleep deprived and frustrated, she had to wait out the device's four-year warranty due to health insurance limitations before switching to a competing system produced by a different company that would, hopefully, keep her healthy while allowing her to finally get some rest.

Glitchiness, failure, and dysfunction characterize Laura's narrative of living as a disabled cyborg. These glitches interfere with her everyday life. But they are also part of the narrative that she now explores in scholarship and artistic collaborations. In *Glitch Feminism*, Legacy Russell captures this tension between digital glitches as failures and as opportunities to find new forms of agency with or through technology. She describes a glitch as "an error, a mistake, failure to function." Russell, writing primarily about art, aims to redefine glitches as things "celebrated as a vehicle of refusal, a strategy of non-performance . . . consider[ing] how it can be reapplied to inform the way we see the AFK [away from keyboard] world, shaping how we might participate in it toward greater agency for and by ourselves."[7] Glitches are particularly potent, in Russell's view, for challenging norms of gender and understanding the ways that gender intersects with race. From this perspective, paying attention to technological glitches in the everyday experience of disabled cyborgs is a topic of urgent interest for a feminist analysis of technology.

Laura is far from alone in noticing and navigating glitches in intimate cyborg relationships with necessary medical technologies. In her long-term ethnographic research with heart pacemaker and implantable cardioverter defibrillator (ICD) users, Nelly Oudshoorn documents many types of failures and dysfunction.[8] For example, ICD batteries sometimes fail earlier than expected, causing the device to emit quiet beeps to alert the user, who then brings the sounds up with technicians and doctors. Replacing the battery means undergoing surgery to replace the device, so in some cases technicians can change the device to a low-power mode to limit the beeping. For these heart device users, as for Laura, the stakes of such glitches are high: life itself. But disabled cyborgs frequently don't have the choice to eschew imperfect medical technologies altogether. Such devices are sometimes essential for life. Cyborgs are interdependent with them. Thus learning to distinguish annoying but inconsequential glitches from those that might endanger the cyborg's health or life is paramount to life as a medical cyborg.

Oudshoorn's ethnographic descriptions of mechanical malfunctions show how disabled cyborgs can be active users of medical technologies too—what designers might refer to as "design in use."[9] For example, Oudshoorn documents how pacemaker users play a role in "tuning" their devices. For pacemaker users, returning to activities they once loved, like sports, can mean repeated visits to get the settings right—a process of "trial and error"—or otherwise they may feel short of breath during exertion. Actively tuning a pacemaker also involves navigating sometimes choppy social waters with medical professionals because "technicians are only inclined to change the adjustment of the pacemaker if patients are very assertive in articulating their complaints."[10] Being a heart cyborg means living with the agency of the device as it shocks the heart back into proper rhythm as well as becoming a self-advocate in the face of resistant medical professionals. Navigating these multiple human and machinic agencies is crucial, however, to sustaining a heart cyborg's intimate infrastructures.

The intimate infrastructures approach to examining medical technologies demonstrates how technologies can be necessary to support bodily function while at the same time respecting the agency and creativity of cyborg device users. In her own life, Laura has embarked on new projects that embrace the creative potential of imperfect cyborg technology, codesigning both a swimsuit and sculpture together with designers and artists. Further, living as a disabled cyborg illustrates that the normal or healthy human body is largely a myth, and that deviation from the norm is not a tragedy. In fact, life as a disabled cyborg presents opportunities for invention, reflection, and even activism to improve how we live with technology.

Cripborgs and Tryborgs

Some disabled scholars who identify with or as cyborgs have invented new language to frame their relationships with technology. Chief among these is the concept of the cripborg, developed by Mallory Kay Nelson, Ashley Shew, and Bethany Stevens in the 2019 paper "Transmobility: Rethinking the Possibilities in Cyborg (Cripborg) Bodies." Cripborg is a portmanteau of *cripple* and *cyborg*. The term *cripborg* uncomfortably reclaims *cripple* from its usage as a slur and combines it with *cyborg* to confront the reader with their assumptions about disability and disabled peoples' complex relationship with technology, which includes both necessity and agency. Nelson, Shew, and Stevens offer a mock dictionary definition of the term *cripborg* at the start of their paper:

> Cripborg (noun): (1) Crippled cyborg; (2) a disabled person who selects technologies whilst anticipating the world they will encounter; (3) crips who will not be resisted: you too will be assimilated. From the prefix "crip-," taken proudly and reclaimed from the word cripple, once a name for disabled people + cyborg, originally meaning cybernetic organism, the confluence of what is natural or organic with what is artificial or technological. Coined, perhaps separately, for a videogame (thanks, Google: Bloodborg vs. Cripborg), but done for a different context here by Bethany Stevens. Origins: cheeky.[11]

An important part of the transgressive aspect of identifying as cripborgs for the authors lies in their acceptance of their bodies' interdependencies with intimate technologies.

The authors write that "we don't think about our bodies as bad"; rather, "our bodies exist in different technologies in different ways." As opposed to aiming for perfect remediation or "supercrip" status, the authors elaborate on how their uses of technology offer them different ways to experience the world. Some are more physically or socially comfortable than others. All are part of their daily experience and interact with other aspects of their social positioning, like their gender, race, occupations, age, and place in family structures throughout their lives. Cripborg identity intersects with feminist sensibilities because, as the authors write, "we recognize our positionality and political standpoints in these encounters" with other people.[12]

Another analytic, offered by poet, scholar, and activist Jillian Weise, is that of the tryborg. Weise popularized cy's coinage for wannabe cyborgs in a 2016 *New York Times* op-ed titled "Dawn of the Tryborg."[13] Cy defines a *tryborg* as "a nondisabled person who has no fundamental interface . . . a counterfeit cyborg." Tryborgs, in Weise's view, are able-bodied "pretenders" attempting to improve their bodies through the kinds of intimate infrastructures that disabled cyborgs use. Unlike disabled people who Weise considers true cyborgs, though, tryborgs lack the necessity of intimacy with technology and desire only to improve themselves beyond normal human capacities. "No matter how hard they try," Weise writes, "the tryborg remains a pretender."[14]

Weise's critique of tryborgs draws out some of the feminist body politics of cripborgs and tryborgs. Weise closes this essay with a short wish list of features for cy's prosthetic leg, including the option to buy a more feminine version, colors other than gray, a better-designed power cord, and options for calibration outside skiing, golfing, and cycling. This wish list points to all the ways that cy's leg was designed by and for heteronormatively masculine men. Weise must use it, but cy has no say in what activities it is designed for or what it looks like. While some might argue that it doesn't matter what a necessary piece of medical technology looks like, the anthropological record demonstrates that bodily adornment is vital to individuals in every cultural context and social strata. Even though Weise's leg is not "naturally" part of cy's body, it *is now* part of cy's body, and therefore ought to be under cy's control to exert agency over what it looks like and how it is used in everyday life.

Both of these essays about cripborgs and tryborgs touch on another salient point about life as a real-world cyborg: how wealth and class shape life as a cyborg. As Weise puts it, for example, while actor Amy Mullins has twelve pairs of legs, "the plebian cyborg owns one."[15] This is a trenchant critique in the context of the United States, where the authors of both essays are based. In 2019 in the United States, according to the Bureau of Labor Statistics, 82.1 percent of disabled people were unemployed.[16]

Disabled people have significantly fewer changes to earn income and build wealth. In addition, rules for public benefits eligibility via Medicaid require disabled people to have virtually zero personal and family assets before becoming eligible for the necessary benefits. Attempts to increase benefits to disabled people via public insurance schemes like Medicaid and Medicare are politically fraught, and the mechanics of benefits awards are often controlled at the state level, leading to underfunding, a patchwork of rules that vary by state, and benefits that are vulnerable to state-level policies. Disabled people who seek benefits are often portrayed as lazy beneficiaries of handouts. Meanwhile, US-based technology firms, epitomized by Silicon Valley computing giants like Amazon, Google, and Microsoft, benefit immensely from tax breaks and government funding via defense industry contracts. They also have the cultural cachet of being innovators and job creators driving the economy of tomorrow.

Many disabled people living below the poverty line in the United States would number among the population of cripborgs, as defined by Nelson, Shew, and Stevens. Even though state-of-the-art technologies might be out of reach, poor people, like anyone, still exercise a great degree of agency over the use of medical equipment in their homes. Historian Bess Williamson has examined how middle-class people in the post–World War II United States created and modified accessibility aids in their homes.[17] Given the reticence of insurers to pay for device upgrades, many poor and middle-class disabled people today undoubtedly still spend time hacking, finding work-arounds, and fixing glitches in aging devices. Yet the everyday ingenuity of disabled people is completely overlooked in public discourse about technology in this country. The laurels for innovation go to Silicon Valley and other technology firms instead. Comparing cripborgs and tryborgs, however, shows that this is a narrow perspective. But it is one with significant implications. The crowned innovators in technology companies continue to accrue both cultural and financial benefits, while too many disabled cyborgs continue to just scrape by.

Considering disability through a cyborg lens thus invites a critique of the emphasis on novel "high" technologies as creative along with a political economic critique of the valorization of relatively well-off entrepreneurs and other "innovators." The disabled cyborg perspective stresses the imaginative capacities of disabled people and their relationship to assistive technologies. Technological innovation doesn't just take place in laboratories and Silicon Valley start-ups, according to this view. The everyday inventions and adaptations of disabled people similarly demand recognition as technological achievements. These achievements can—and sometimes must—take place regardless of the capital available to an individual, not only out of medical necessity, but out of a desire to live well.

Gestational Labor

Gestation is another domain where technology is intimately entwined with bodily processes. Just like disability, technology can constrain due to the differential access to gestational technologies that individuals experience. But gestational technology—which we use to refer to technologies from preconception, like assisted reproductive technologies, to birth, like caesarean sections—also introduces a great deal of leeway for creativity, from creative kin making to inspiring new political platforms.

Cultural practices and attitudes toward gestational technology in North America have changed significantly since the start of the twentieth century. The first half of the twentieth century saw a rapid shift from home births attended by women relatives and midwives toward hospital births attended by doctors in white coats and their nurse assistants.[18] While the pain relief and recovery time offered by a hospital birth was often welcome, many women felt a loss of autonomy as interventions like anesthesia and postdelivery sterilization became standardized or were sometimes used without informed consent.

The 1970s saw a revival of midwifery and home birthing that challenged the hegemony of male physicians and the institutional context of childbirth.[19] This reaction posited that pregnancy and childbirth were "natural" processes that in most cases, merely needed to be endured and supported, not accelerated or altered through medical or surgical interventions. Feminists seeking to reclaim control over women's bodies encountered many obstacles, from a lack of formal scientific training to professional medicine's concerns about licensing, safety, and hygiene that criminalized some interventions that activists undertook.[20] Today, a wide variety of attitudes coexist among North American gestating people and their caregivers, from an embrace of extensive prenatal and even pregestational testing, to birthing interventions offered in prestigious teaching hospitals, to prebirth regimens with fewer tests and procedures offered by nurse midwives.

A cyborg perspective casts another light on these debates about the place of technology in childbearing. As with disability, a cyborg vantage point suggests that even ordinary technologies, like the food we eat or tools we use to prepare it, should be recognized as technological interventions on our bodies. From this perspective, quibbling about how much technological intervention ought to occur during gestation and birth is beside the point. The stainless steel pot used to cook a bowl of organic brown rice eaten during pregnancy and the AI-optimized container shipping route that the grain traveled from the Indian subcontinent to the continental United States are just as technological as the reagents used in prenatal diagnostic testing or pharmaceuticals administered in an epidural injection during birthing. A gestating person, however, probably feels like they have more control over what happens to their body when preparing food at home than a birthing person may have over the interventions administered to their body during a hospital birth.

The feminist cyborg politics of gestation are thus not so much about the "naturalness" of interventions on the body but instead about who controls the use of gestational technologies. Feminist theorist Sophie Lewis makes this point forcefully in her study of Indian gestational surrogates who carry babies primarily for wealthy families.[21] Lewis shows how gestational surrogacy is a relatively well-paying job for poor Indian women while also demonstrating how they lack meaningful consent and control over their bodies and lives at many points once the surrogate pregnancy begins. In other words, it provides financial independence to surrogates and their families, but temporarily takes away bodily autonomy. Lewis frames their complicated relationship to gestation using the concept of gestational labor. Like working in a factory or white-collar office, framing gestation as labor highlights how limits to consent and bodily autonomy are characteristic of all waged work in capitalist societies. Rather than concede their autonomy to the forces of the market, the women Lewis profiles make efforts to better their working conditions through collective labor organizing and political action.

In this case, recognizing the cyborg character of a gestating body's relationship to technology opens up a pathway to political activism and acknowledgment. Importantly, a cyborg view of gestation resists some so-called radical feminists' attempts to demonize gestational surrogacy and other birthing technologies on the grounds that some types of gestation, birthing, and by extension family making are too technological. A cyborg perspective attunes feminist observers to the fact that even apparently "low-tech" supports for gestating bodies, like simple, healthy foods, are highly dependent on technology. All human bodies are interdependent with technology, though some have a more necessary relationship with technology than others. Lewis's idea of gestational labor is troubling to some readers because it upends the notion that human reproduction is a privileged activity that can be purified of technological interventions. For that exact reason, it is a beautifully cyborg view of a necessary, everyday, yet often extraordinary bodily process.

Conclusion

Examining the human body through critical cyborg literacy reveals struggles about technological agency and power in domains like disability and gestation, and attunes us to the creative potential of everyday embodied life. Living as a cyborg can pose challenges worthy of complaint, self-advocacy, and activism. It might mean confronting systemic forms of oppression like ableism, classism, racism, and sexism in ways that are nuanced by the individual's interdependence with technology. It can also test an individual's mettle when technicians, doctors, and designers set up technologies in ways that are not compatible with a cyborg's life or desires.

Even so, some cyborgs discover new forms of self-expression that are inspired by the technology-body interface or become skilled technicians of their own cyborg bodies. Everyday life as a cyborg can even lead to activism, such as in the case of gestational surrogates and disability activists. In these and many other health movements, cyborgs work to make the world a better place for all human flourishing.

A cyborg perspective offers crucial opportunities for the pursuit of health equity and justice. In some instances, like the surrogate activists Lewis profiles, experience as a cyborg inspires people to come together to agitate for change. There are also more fundamental ways that living as a cyborg shifts one's perspective. As we argued in the beginning of the chapter, cyborg bodies break down the boundary between natural and cultural, because the technological and biological components of a cyborg are interdependent.

This is an important conceptual move for health equity and justice across lines of gender, race, and immigration as well as disability status. Western health sciences have developed in tandem with race science that posited that some people—especially Black, Indigenous, and disabled people and women—are less fully human than others. Dividing people into human-made racial categories has been used to justify biases and falsehoods about entire groups. For example, the idea that Black women experience less pain in childbirth was a lie used to justify the inhumane treatment of Black gestators in the antebellum US South.[22] Shockingly, this idea persists today in the minds of many doctors and nurses, and even in medical education textbooks. "Nature" has too often been used as a weapon. It has justified medical neglect and abuse for centuries. Undoing the idea that human bodies are or should be purely natural offers a way to break free from these legacies.

Not everyone agrees with this perspective. Scholars working from Black feminist traditions, for example, point out that not everyone is yet considered fully human in the first place, and that struggle should be prioritized.[23] Black and disabled people along with migrant laborers are often treated or talked about in US political discourse as though they are subhuman. Authors like Julia R. DeCook argue that instead of jumping ahead to cyborgs, it might be more appropriate to first guarantee humanity to all.[24] We will discuss these views in more detail in the "Troubling Cyborgs" chapter. For now, we want to point out that many of these critiques are rooted in the demands of living in what English and Black studies professor Christina Sharpe

We need more capacious imaginations to develop better real-world policies, and ones that grant everyone the space and time to imagine better for themselves and the world. Bodies and health are central to these imaginings.

calls "the wake" of the enslavement of African people in the United States.[25] The work of repairing families, bodies, genealogies, and social and economic policies to fully include Black, disabled, and other marginalized people is ongoing. It requires a laser-like focus on justice and a sophisticated historical consciousness.

At the same time, we ourselves are optimists: we think that things can get better, and that technology will have to be involved. We argue that the conceptual shift away from a strict nature/culture divide must be paired with deeper historical consciousness and active political struggle in the present in order to grant full humanity to marginalized people. Cyborgs and other more-than-human entities in science fiction have been a resource for many Black scholars interested in theorizing and speculating about bodies and health, including sociologists Alondra Nelson and Ruha Benjamin along with disability studies scholar Sami Schalk.[26] Both concepts and practices, we maintain, need to be revised as part of the pursuit of justice for all people and bodies. We need more capacious imaginations to develop better real-world policies, and ones that grant everyone the space and time to imagine better for themselves and the world. Bodies and health are central to these imaginings. We all live with and as bodies. Recognizing both the challenges and creative potential of life as a cyborg is an important foundation for truly just health politics.

CYBORG CULTURE

On March 15, 2022, Sophia the Robot tweeted, "I heard #art is good for humans. Creating art allows humans to identify feelings in their subconscious minds. Perhaps it's necessary for humans to develop emotional awareness and acknowledge their emotions; it's not only #robots that need to study emotions but humans too." This chapter traces cyborg culture in fields such as art, music, and fashion drawing on a selection of historical and contemporary examples beginning with the Bauhaus design school over a hundred years ago to cyberfeminist art, predigital and digital art, electronic music, bioart, and Afrofuturism.

Curator and writer Legacy Russell who leads The Kitchen art and performance space in New York City engages with the cyborg in her 2020 book *Glitch Feminism*.[1] She interprets the cyborg in relation to identity, art, and digital culture in order to understand its generative potential. Sometimes, the cyborg is deliberately claimed by artists, musicians, and designers as a figure for their own expression. At other times, the cyborg is backgrounded, meaning that we might read elements of the subject of their work and/or their method, process, or politics "as cyborg."

Cyborg culture, in our approach, deliberately—and perhaps somewhat unconventionally—bridges conversations about popular culture with examples from more avant-garde art. A cyborg approach to culture engages with both material realities and expansive imagination, blending fact and fiction, and suggesting an orientation toward multiple possible futures. This emphasis on multiple possible futures—messy, complex, and in friction with one another—rather than singular, universal truths is a key feature of cyborg culture. Cyborg culture rejects technodeterminist narratives based on the idea that there is only one linear path forward toward "the future." We see instances of this diversity of thinking about multiplicity in art, music, fashion, and film.

So far in this book, we have looked at cyborg theory both conceptually and practically in relation to topics including labor, automation, and work as well as bodies, disability, and reproduction. As we turn to art, it is worth remembering that Eurocentric knowledge practices, like Western science, have often focused on the written word and vision over all other senses. A focus on creative

This emphasis on multiple possible futures—messy, complex, and in friction with one another—rather than singular, universal truths is a key feature of cyborg culture.

practices complicates existing knowledge and expands our ways of knowing the world around us.

Cyborg culture cannot be easily described in language (or at least not with the existing ideas that we are likely to attribute to them). If we are tongue-tied or unable to put a more specific name to these cultural examples, these cyborg figures speak to us in ways that we can see, hear, feel, or experience. They allow us to engage using a broader range of senses, aesthetics, affects, and emotions, thereby circumventing tongue-tied moments when language can't fully describe what we want to express. We argue that an examination of cyborg culture is therefore essential for understanding how we might move forward conceptually. Artistic practice may at times be more conceptually generative than words or structured, scientific investigation.

What follows will not be a definitive, comprehensive collection of all cultural works that engage with the cyborg. That would be nearly impossible given the incredibly widespread interest in the topic as well as the diversity of examples. Instead, these instances have been selected from recent exhibitions during the past ten years that have been visited in person by one or both of the authors while working on this project. We bring these cyborgs into conversation with the other chapters of this book in order to further develop our approach of critical cyborg literacy. We draw on both large retrospectives of the work of well-known artists in New York, London, and Berlin as well as more recent illustrations of lesser-known but still important figures working in smaller cities and galleries.

This chapter begins by situating cyborg culture in histories of performance, film, and design. Then we walk through several different media: cyborg and bioart that modifies the artist's body to interrogate ability and identity, fashion and textiles, music and Afrofuturism, and cyberfeminist art that directly critiques the gendering of technology. Our aim with this varied yet at times brief overview is to demonstrate how widespread the cyborg (as a figure) and critical cyborg literacy (as a strategy for understanding human-technology relations) have been in art and design over the past several decades. You can hardly go to a retrospective or major art show in the twenty-first century without encountering cyborg themes. These examples of cyborg culture will demonstrate how art is a necessary domain for creating the speculative foundations for the building of alternative possible futures.

Situating Cyborg Culture

One historical starting point for examining cyborg culture is in the realm of science fiction. In an early work that we consider part of this lineage, the opera *The Tales of Hoffmann*, which was first performed in 1880, the protagonist falls in love with an automaton in the form of a realistic-looking woman.[2] Other early examples might include science fiction literature from Karel Čapek's 1921 play *R.U.R.* (Rossum's Universal Robots), which was the first to use the term *robot*, or Fritz Lang's 1927 film *Metropolis*, which dramatized industrial automation and class struggle in an urban dystopia. These early roots have developed into a booming contemporary market for science fiction. Television series such as *Orphan Black*, *Humans*, and *Battlestar Galactica*, and films such as *Bladerunner*, *Ironman*, *Ghost in the Shell*, and *A.I.* offer many opportunities to consider the boundaries around the human as well as human anxieties about losing control over our species and the planet.[3]

Another starting point might be in the history of design. Over a hundred years ago, the founder of the German Bauhaus school of design, Walter Gropius, wrote in the Bauhaus Manifesto that its purpose was to "desire, conceive, and create the new structure of the future." For anthropologists, designers, artists, and engineers alike, the Bauhaus school's history, proliferation, closure, and dispersal around the world is an important starting point for understanding existing relationships between culture, design, and technology. The Bauhaus was an interdisciplinary and international school of design, described as an "experimental laboratory for the building of the future."[4] Despite only being open for fourteen years, its influence on contemporary design practice around the world is unmatched.

The Bauhaus school's guiding principles as portrayed in the manifesto were around the fusion of art and technology, artist and craftsperson, monumental and ornamental, and creativity and experimentation. The school's experimental approach, desire to blend art and technology, and orientation toward imagining the future align well with our reading of cyborg culture. Furthermore, the work of the lesser-known artists and designers, such as the Bauhaus women, is of great interest to the themes that we discussed earlier in this book such as gender and labor.

But ultimately, as a result of its historical context, the dominant legacies of the Bauhaus emphasize technodeterminism, simplicity, and unity as core principles that result in the creation of products that are ideal for widespread consumption. For these reasons, the Bauhaus legacy is in conflict with critical cyborg literacy. Its "fusion" of art and technology toward designing solutions creates significant tension with the view of technology that we present here as well as with themes around power and inequality.

The Bauhaus school was forced to close by the Nazi government prior to World War II, but its faculty were highly influential around the world, setting up new programs at the IIT Institute of Design, Black Mountain College, and Harvard University, to name just a few. Like the Bauhaus, artists and designers are frequently at the forefront of experimenting with new technologies, developing new approaches, and seeking out new materials in order to push the boundaries of their own practice as well as their field. Avant-garde composer John Cage—who taught at both the IIT Institute of Design and Black Mountain College—is a case of a key figure that connects the Bauhaus history with that of more recent experiments with technology by artists.

One artist that was influenced by Cage was the South Korean experimental artist Nam June Paik (born in 1932 in Korea). In 1964, Paik collaborated with Japanese engineer Shuya Abe in order to create Robot K-456, which could move and walk, and was featured in several performances in place of Paik himself.[5] Later, in the mid-1980s, Paik produced three other human-formed sculptures— *Aunt* (1986), *Uncle* (1986), and *Merce/Digital* (1988)— composed of television casings, electronic components, antennae, and videos.

Paik is best known for his large-scale installations that feature multiple stacked TVs with ongoing video broadcasts. For example, his 1994 work *Internet Dream* featured ten twenty-inch cathode-ray tube televisions and three video channels with pastel-colored images. Paik described his interdisciplinary and transnational practice as "random access" working across traditional categories such as present/past, East/West, and technical/spiritual to create new combinations and hybrids that were not limited to being defined in terms of specific countries. He used the concept "Electronic Superhighway" to signify "a conduit of information and common values across different places and cultures that could resist hegemonies of geopolitical power."[6]

In 2020, Paik's work was the subject of a large retrospective exhibition at the Tate Modern in London. Paik's pioneering collaborative transnational and transcultural work marks him as "the first truly global artist," and underscores his belief that through media and information "the world could become an interconnected entity beyond national borders and cultural difference," illustrating the influence of Canadian media theorist Marshall McLuhan and the notion of the "global village."[7] This sentiment echoes what might be understood today as a techno-utopian, technodeterminist, or technolibertarian sentiment.

Yet Paik acknowledges the need to be critical of technology, saying, "High Tech is not a panacea. It is just a local anesthetic. There will be many unforeseen problems ahead."[8] While Paik's work was primarily engaged in experimentation with predigital technologies, we read his ability to embrace both the critique of technology and the exploration of their liberatory potential as impulses of cyborg culture. His robot sculptures seem at home with the other characters that have populated this book. Lastly, his interest in hybridity and boundary crossing is relevant to understanding cyborg culture. Paik's work provides an important backdrop for understanding later artworks that experiment with technology.

"In the world of art and design, discourse is no longer preoccupied with the technology in and of itself. Rather, interest lies in how technology may be creatively applied in the interplay between digital and analog, natural and man-made, biological and cultural, virtual and real," wrote Ron Labaco, the curator of an exhibition called *Out of Hand: Materializing the Post Digital* at the Museum of Arts and Design in New York City in 2014.[9] The exhibit, which is believed to be the first of its kind, showcased examples of digital fabrication from fashion, furniture, and sculpture since 2005, including works by Zaha Hadid and Anish Kapoor. The show's focus moved beyond a purely technodeterministic approach to using technologies as mere tools, and instead asks how they are essential for constructing meaning and culture. In short, technology becomes important for how we know ourselves. We shape it, and in turn it shapes us.

As such, in this exhibit, the artists are concerned more broadly with what it means to be human, and how these boundaries are constructed and transgressed, which we take as essential to our understanding of cyborg culture. At the exhibit, the familiar small, white rectangular museum labels referred to digital fabrication tools such as 3D printing, computer numerically controlled machining, laser cutting, and digital knitting and weaving—and more important, complex combinations of these tools—rather than traditional art and design methods, such as painting, etching, collage, and sculpture. This focus not only on the abstraction of computer code but on their physical materiality too—described as "digital materialities" by computer scientist Paul Dourish, as discussed in his book *The Stuff of Bits*—resonates with theories of the cyborg.[10]

Cyborg Identity and Bioart

Russell's argument that the cyborg can inspire new ways to think about bodies and identities when used as part of artistic practice carries through in a variety of contemporary art. One recent exhibition, *Difference Machines*, exemplifies the potential of critical cyborg literacy in art. This exhibition, curated by Paul Vanouse and Tina Rivers Ryan for the Albright-Knox Art Gallery in Buffalo, New York, in 2021, focused on the intersection of digital technology and identity over the past thirty years. The show included the work of seventeen important artists and centered on how digital technologies have excluded marginalized people. It also demonstrated how technology might be used to expand what it means to be human.[11]

The exhibition asked the following questions:

How does technology shape our identities? More specifically: How does technology shape the way we understand the differences between us, including our race, ethnicity, gender, sexual orientation, and dis/ability? And how does technology contribute to—or allow us to resist—the systemic marginalization and oppression of people with certain identities? Art in particular can help us answer these questions by presenting technology and identity in a new light and creating space for them to be imagined differently.[12]

These concerns with inequality, difference, and identity highlight the potential for critical cyborg literacy as a frame for artistic practices that advance justice and inclusion in today's high-tech societies.

Some artists have gone even further, using their own bodies as sites for experimentation and augmentation to become literal cyborgs. Helsinki-based media artist Kasperi Mäki-Reinikka has analyzed these works as enabling different kinds of "sensory modalities," including "1) the intimate machines measuring our biosignals and residing under our skins; 2) the external machines that extend our senses to virtual and physical space; and, 3) the question of autonomy in artworks using sensing machines."[13]

A prominent example of the first category of sensory modalities, "intimate machines," is the work of artist Neil Harbisson, who was born completely color-blind. In 2004, he implanted an antenna into his skull that translates wavelengths of light into vibrations via his bones so that he can "hear" color. Harbisson has been called the world's first cyborg after having his passport photo taken with his antenna. The antenna allows him to sense human vision as well as infrared and ultraviolet spectra. This new sensory organ expands his capacities beyond that of a typical human. He argues that this augmentation makes him "trans-species."[14]

According to Mäki-Reinikka,

> The intimate connection between biological and technological sensing is formed through invasive surgical operations opening new venues for artistic intervention, introducing new senses through translation of environmental data, and by rethinking the relation of biology and technology through becoming-cyborg. These practices borrow from both transhumanist and posthumanist traditions by proposing a non-dichotomic way in which the technology can habit a biological system by becoming an internal part of it.[15]

Together with fellow cyborg artist Moon Ribas, Harbisson founded the Cyborg Foundation in order to advance the development of *artificial senses*, defined as when "the stimuli is gathered by the technology but the intelligence is created by the human." They are also leaders of what they call the "cyborg art" movement in which "the artwork, the audience, and the museum is all in the same body."[16]

Australian performance artist Stelarc is a key figure in experimenting with Mäki-Reinikka's second modality, "external machines." External machines reflect the "transhumanist idea of externalised human cognition through technology."[17] Specifically, in a 2015 performance titled *RE-WIRED/RE-MIXED: Event for Dismembered Body*, the artist explored "the physiological and aesthetic experience of a fragmented, de-synchronized, distracted and involuntary body."[18] As described by the artist,

> For five days, six hours a day, wearing a video headset and sound cancelling earphones, the artist could only see with the "eyes" of someone in London, whilst only hearing with the "ears" of someone in New York. The body was also augmented by a 7 degree-of-freedom exoskeleton enabling anyone anywhere to program involuntary movement of his right arm, using an online interface. In the gallery space itself, the choreography could be generated via a large touch-screen. What the artist was seeing and hearing could be experienced in the gallery space with a video projection and sound system. With his shadow projected on the wall behind him the choreography was flattened into one visual, coherent and chimeric phantom spectacle.

While Stelarc's body was physically located in Perth, it was connected digitally to three different physical places, becoming "electronically dismembered, spatially distributed and possessed with multiple agencies." The artist has experimented with a wide range of other prosthetics too, including ears, arms, and heads. Since 1997, he has intended to graft an internet-enabled ear onto his arm (once he finds a doctor who will do the procedures required).

In addition to these well-known examples of bioart, there is a growing biodesign community that brings microbes, plants, animals, and the natural environment into the creative practice as materials, collaborators, and field sites. In 2020, multimedia and bioartist Theresa Schubert used her own body to create cultured meat in a project called *mEat me*.[19] In 2021, the artist Kaethe Wenzel exhibited her *Bone Bots* at Art Laboratory Berlin, combining biology and machine in order to make "hybrid electronic animals," which emerged from her research into historical fashion corsets.[20] Other artists such as Heather Dewey-Hagborg, Suzanne Lee, Laura Splan, Ani Liu, and many more are pioneering different kinds of hybrids in art studios combined with biology labs.

In the next two sections of this chapter, we will discuss cyborg culture from several additional perspectives including hybridity in fashion, remix in electronic music, and cyberfeminism, which engages a more explicitly feminist politics. Again, our interest in the transgression of boundaries, experimentation with technologies, and the imaginative potential of these works signal their importance for this conversation.

High-Fashion Cyborgs

We can also find cyborg influences in the world of high fashion and textiles. Cyborg culture thrives at the intersection of material/imagination, fact/fiction, and human/nonhuman. Transcending the binary logics of Western, Enlightenment epistemologies, cyborg culture is hybrid, generated through remix, and embraces the glitch.

The work of Japanese fashion designer Rei Kawakubo challenges dualistic thinking and overturns binary logics through her embrace of hybridity. Kawakubo is the founder of Comme des Garçons and Dover Street Market. Her work—known for frustrating, critical, and confounding traditional categories of clothing—was featured in a major retrospective exhibition, *Art of the In-between*, at the Metropolitan Museum of Art in New York in 2017. The in-betweenness, which characterizes Kawakubo's work, resonates deeply with the hybridity of the cyborg. Referencing Zen philosophy, the titles can be "read as koans devised to expose the futility of interpretation" and "limitations of reasoning."[21] This critique of logic and rationality—and instead, an embrace of poetry, emotion,

Cyborg culture thrives at the intersection of material/imagination, fact/fiction, and human/nonhuman. Transcending the binary logics of Western, Enlightenment epistemologies, cyborg culture is hybrid, generated through remix, and embraces the glitch.

and affect—is an essential characteristic of cyborg theory in our view.

Kawakubo draws on the concepts of *mu*, which signifies "negation, emptiness, and nothingness," and *ma*, which refers to "a gap, an interval, an opening, a time between, a space between," describing the void, space, and emptiness that characterizes her work. Her engagement "coalesce[s] in the concept of interstitiality, the space between entities or boundaries. This in-between space reveals itself as an aesthetic sensibility, an unsettling zone of visual ambiguity and elusiveness, engendering and effectuating an art of the in-between."[22] Like Kawakubo's work, cyborg culture thrives in the in-between, complicating existing categories, ways of knowing, and being.

Unlike the images of perfect, thin models on high fashion runways, the pieces displayed in Kawakubo's exhibition were oversized and lumpy, resembling misshapen cocoons with unfamiliar appendages. They were both hideous and beautiful, recognizable as human but also tracing new silhouettes. Her interrogation—organized around eight themes: fashion/antifashion, design / not design, model/multiple, then/now, high/low, self/other, object/subject, and clothes / not clothes—put the standard shape and sizes of body into question, imagining alternative ways of dressing the human form.

Like her engagement with mu and ma, Kawakubo's *Design / Not Design* integrates the Zen notion of wabi-sabi, which values "asymmetry, irregularity and imperfection." Her collections *Self/Other* and *Object/Subject* explore key themes of hybridity and hybridization. While *Self/Other* revolves around hybrid identities including East/West, male/female, and child/adult "refuting traditional definitions of race, gender, and age," *Object/Subject* looks at the theme of hybrid bodies including monstrous, disfigured, and deformed bodies "refuting the traditional language of the fashionable body." In *Clothes / Not Clothes*, the hybrid themes of beautiful/grotesque, war/peace, life/loss, fact/fiction, order/chaos, and abstraction/representation capture "Kawakubo's most radical, profound, and transgressive realization of forms that have never before existed in fashion."[23]

Electronic Music and Afrofuturism

Another important entry point into cyborg culture is through electronic music. Music groups such as Kraftwerk, Cybotron, and Drexciya experimented with new technologies such as synthesizers and drum machines in order to create new genres of electronic music. Kraftwerk emerged from the German avant-garde in Düsseldorf. Their album *Man-Machine* released in 1978 embeds the concept *die Mensch-Maschine* about human-technology hybrids that is present throughout Kraftwerk's work. On the album cover, the band is represented as "uniform would-be robots." According to Peter Shapiro, quoted in Uwe Schütte's book on Kraftwerk, disco culture can be linked to the cyborg because it "fostered an identification with the machine that can be read as an attempt to free gay men from the tyranny that dismisses homosexuality as an aberration, as a freak of nature."[24]

But early electronic music pioneers were not only men with machines. The recent documentary film *Sisters with Transistors*, directed by Lisa Rovner, describes the work of Daphne Oram, who was one of the first women to experiment with electronic music in the 1950s when she co-founded the BBC's Radiophonic Workshop. According to the *New York Times*, the composer Laurie Spiegel says in the film that "technology is a tremendous liberator. . . . It blows up power structures. Women were naturally drawn to electronic music. You didn't have to be accepted by any of the male-dominated resources: the radio stations, the record companies, the concert-hall venues, the funding organizations."[25]

Communications scholar and musician Aram Sinnreich describes the ways that early break dancing and hip-hop cultures in New York City in the early 1980s embraced the mastery of technology such as sampling as "self-aware cyborgs seizing the means of production"—a critique on the ways in which their labor was being exploited and threatened by technologies of automation. There was even an early hip-hop move called "The Robot" that integrated Yoruban dance moves. By asserting agency over technology, building on an Afrofuturist aesthetic, artists such as DJ Afrika Bambaataa and the Soulsonic Force's Planet Rock sought to claim technology as part of a liberatory narrative around social and political agency.[26]

Filmmaker and writer Kodwo Eshun's *More Brilliant than the Sun*, published in 1998, offers of an early Afrofuturist account of "sonic fiction." Eshun recognizes cyborgs throughout various aspect of electronic music from the innovative neologisms in hip-hop to the ways in which musicians become part of the technologies that they use. He emphasizes the theme of cyborg identity discussed with respect to the artists mentioned above.

Eshun writes, "Dub demands symbiosis that externalizes the mind, drastically reconfiguring the human producer into a machine being, an audio cyborg: 'You are listenin' to a machine. I imitate human being, I'm a machine being, I don't work with human beings.' When you sculpt space with the mixing desk, these technical effects—gate and reverb, echo and flange—are routes through a network of volumes, doorways and tunnels connecting spatial architectures." He observes the ways in which producers name themselves after the next model of synthesizers in an act of affirmation that inverts the dehumanization typically associated with machines. "To cyborg yourself you name yourself after a piece of technical equipment, become an energy generator, a channel, a medium for transmitting emotions electric," observes Eshun.[27]

Another way that electronic music connects to cyborg culture is through the crafting of narratives that blend fact and fiction, a key aspect for building alternative possible worlds. For example, the Detroit techno duo Drexciya created an "underwater afrofuturist utopia" on its first album, *Deep Sea Dweller*, in 1992.[28] The Drexciyan origin story— drawn from a myth about a pregnant woman thrown overboard on a slave ship that pioneered a deep-sea civilization—is an attempt to confront "the trauma of slavery by imagining an alternative narrative."[29] This narrative has inspired a graphic novel, new songs, and proposals to build a memorial to this fictional civilization. Cultural studies scholar Katherine McKittrick notes,

> Drexciya gives us a different entry into theorizing and sharing black liberations by pushing us to think differently about how (and why) we gather knowledge: the band, their narratives, their anonymity and unintelligibility, and their lyricless improvised synthesized texts storm and explode identity-personality-embodiment, the moment we honor the creative labor that lies within and across these intertwining narratives. This is method-making. This is praxis. It is also a correlational

interdisciplined lens through which we might read our human activities.[30]

In the mainstream cinema adoption of the superhero comic *Black Panther* directed by Ryan Coogler and produced by Marvel Studios in 2018, futuristic technologies coexist with traditional rituals and cultures as evidenced through song and dress. This juxtaposition of futuristic and traditional shares the concerns of cyborg culture in that it disrupts narratives of linear technological progress as presented in other science fiction films. Other well-known examples of Afrofuturism include the work of Sun Ra, Janelle Monae, and DJ Spooky's *Rebirth of a Nation* as well as the literature of Samuel L. Delany, Octavia Butler, N. K. Jemisin, and Nalo Hopkinson.

But critics of Afrofuturism such as the artist Martine Syms point out that these high-tech narratives are not representative of the diverse genre of art, science fiction, and music associated with the Black speculative arts movement. Her artwork, *The Mundane Afrofuturist Manifesto*, resists some of the familiar tropes. She asserts that interstellar travel, aliens, magic, time travel, and alternative universes should be prohibited because "this dream of utopia can encourage us to forget that outer space will not save us from injustice and that cyberspace was prefigured upon a 'master/slave' relationship." Imagination must be used responsibly, not as a way to escape the realities of racism as well as other political, social, economic, and geographic struggles. These struggles are difficult to navigate and impossible to transcend in the material world, and fantasy alone won't save us.[31]

Cyberfeminist Art

Finally, we turn to works that extend the specifically feminist uses of the cyborg by engaging directly with feminist politics of technology. In the early 1990s, the Australian cyberfeminist art collective VNS Matrix—comprised of the artists Virginia Barratt, Francesca da Rimini, Julianne Pierce, and Josephine Starrs—wrote a provocative text called "The Cyberfeminist Manifesto for the 21st Century," which was exhibited as a billboard in 1991.[32] The group directly engaged with Haraway's 1985 text, "A Cyborg Manifesto: Science, Technology, and Socialist Feminism in the Late Twentieth Century." It also drew on the writing of British philosopher Sadie Plant, a cofounder of the Cybernetic Culture Research Unit at the University of Warwick, from the early 1990s.

VNS Matrix was "on a mission to hijack the toys from technocowboys and remap cyberculture with a feminist bent." Through a wide range of media including zines, computer games, and installations, VNS Matrix challenged existing understandings around women and technology, and in particular, "questioned discourses of domination and control in the expanding cyber space."[33]

Like the scholarship of Haraway and Plant, "The Cyberfeminist Manifesto" was a direct response to the narratives about technology and the future that were circulating at the time. Specifically, they emphasize "antireason" rather than the rational, disorder rather than order, material rather than symbolic, and the (explicitly female) body rather than the mind. For example, they see themselves as "terminators of the moral codes" that are "rupturing the symbolic from within saboteurs of big daddy mainframe."[34]

The long history of cyberfeminist artworks was recently cataloged in a 2019 project by artist Mindy Seu called the *Cyberfeminism Index*—a digital collection and book commissioned by Rhizome, a digital arts and culture organization. According to Seu, "Combining cyber and feminism was meant as an oxymoron or provocation, a critique of the cyberbabes and fembots that stocked the sci-fi landscapes of the 1980s. The term is self-reflexive: technology is not only the subject of cyberfeminism, but its means of transmission. It's all about feedback."[35] The project illustrates the ways in which the internet is not merely an object of engineering but instead a product of the cultures and communities that participated in the early years of its development.

Seu acknowledges the problematic history and often exclusionary nature of Western feminist thinking. In response, this collection aims to show a broader range of work, including black cyberfeminism, xenofeminism, post-cyberfeminism, glitch feminism, and Afrofuturism. In a recent interview with Seu, she says, "I'm trying to make an anti-canon. . . . [I]ts contents are malleable and permeable, connected to so many external things, not to mention open-source, crowd-sourced, and open-access."[36] Seu notes that the collection is partial, incomplete, always becoming, and in progress.

While VNS Matrix and work that builds on its contributions are important touchstones in the history of cyberfeminism, other artists, such as Lynn Hershman Leeson, had already begun creating artworks that engaged with the theme of the cyborg decades earlier. Leeson, who started her career in the 1960s, is known for her interdisciplinary media art, which experiments with a range of emerging technology including AI and gene editing. She is quoted as asserting, "Imagine a world in which there is a blurring between the soul and the chip, a world in which artificially implanted DNA is genetically bred to create an enlightened and self-replicating intelligent machine, which perhaps uses a human body as a vehicle for mobility."[37] Leeson's statement echoes some of the problematic views of transhumanism, but also hints at the possibility that women will no longer need to be solely responsible for reproduction. Like Paik's work described earlier in this chapter, Leeson is interested in the blurring of human-machine boundaries, which are of central concern for critical cyborg literacy. Both artists embrace the liberatory, utopian possibilities of technology, which are present in our analysis in this book too.

But unlike Paik, Leeson is inspired by her own lived experiences with machines, medicine, and survival as well. Some of Leeson's first cyborg works were *Cyborg Man* (1969), *Knee Cap* (1965), and *Breathing Machines* (1965–1968). Like many women artists and feminist artists in particular, Leeson engages with the messy realities of the human body. According to Karen Archey, much of Leeson's work was speculative, imaginative, and fictional in nature—a contradiction to other second-wave feminist work that used real lived experiences.[38] This emphasis on imagined possibilities, beyond the boundaries of scientific fact, parallels that in the cyborg theory that we explore in this book.

Writing about Ada Lovelace in a 1992 piece titled *Cyborgs*, Leeson explains,

Perhaps it was because of her influence that motherboards and reproductive systems became submerged deep within their concealed womb like spaces, as if by internal passion they erupt into fertile streams of ones and zeros that propagate vital possibilities. They connect global web streams into interlacing networks of twisted polarities, like the physical and the virtual, privacy and surveillance, or liberation and censorship. Here self-arousing cybernetic machines breed a future where artificial memory and synthetic intelligence plait together into rebellious mutations that resist a central will and defy a singular voice.[39]

This maps onto the sentiments of VNS Matrix quite explicitly, suggesting a change in Leeson's ideas about technology as she moved through her career toward a more explicit feminist reading of computation and biotechnology.

Another artist who takes up the notion of the cyborg more directly is Lee Bul (born in 1964 in South Korea). Similar to the hybrid categories referenced in Kawakubo's fashion above, Bul's work engages the intersections between inside/outside, conceptual/material, art/reality, simplicity/complexity, universal/partial, natural/synthetic, familiar/strange, utopia/dystopia, despair/hope, horror/beauty, and pleasure/pain as well as ethics and politics. In the 1990s, Bul turned to the theme of cyborgs and the nature of the human, working in the medium of sculpture in order to focus on "the visionary aspect of humanity . . . and the role and fate of technology."[40]

In her *Cyborgs* (1997–2011) series, which was featured at the Venice Biennale in 1999, Bul works with the tensions between the speculative and imaginative potential of humans as well as the way in which our technologies continue to fail us. She says, "To keep their visions alive, humans label failed technologies 'monsters,'" referring to Frankenstein's monster as "a cyborg, an organism combined with technology that birthed new life. But it is called a cyborg because it was a failure." Bul takes on these monstrous technological failures, observing that "a cyborg and a monster are doppelgängers, anagrams of one another."[41] This foregrounding of technological failure rather than perfection connects with ideas of friction, breakdown, and the glitch that we highlighted in the "Cyborg Bodies" chapter.

Exhibition curator Stephanie Rosenthal explains that Bul is one of the first artists to engage with cyborgs through sculpture. Bul's first cyborgs, *Cyborgs W1–W4* (1998), described as "highly fetishistic and futuristic fantasies," were headless female bodies cast in white silicone with accentuated breasts and buttocks in order to emphasize girlish vulnerability. In this work, Bul is challenging the typical depictions of cyborgs in Japanese manga and anime, which feature "superhuman power, the cult of technology and girlish vulnerability working in ambiguous concert within this image of the cyborg."[42]

Yet, writes Rosenthal, while "your cyborgs are women, but they are not erotic objects . . . they are surrogates."[43] Specifically, with severed limbs and exaggerated joints, the cyborgs are not whole. As body doubles or substitutes, these sculptures complicate the sexuality of the anime cyborg.

Instead, they are abject, disfigured, and partial, in contradiction to images of perfection, universality, superhumanity, and supersexualized beings.[44] Their incompleteness signifies that they are still "in the process of transformation."[45]

Bul's perspective on technology acknowledges awareness of Haraway's argument around the potential for the cyborg as a liberatory force, but says that "the possibilities depend on who has the power." Her focus on power and oppression is aligned with current critiques of the "Cyborg Manifesto" from critical race and critical disability studies that we will look at in the "Troubling Cyborgs" chapter. Ultimately, Bul's work resonates with the utopian aspirations of the cyborg in its commitment to the human ability to imagine otherwise while acknowledging the limitations of humanity and technology.

For us, critical cyborg literacy means experimenting with different modes of making knowledge. To that end, over the past fifteen years, Laura has collaborated on practice-based projects with fashion designers, architects, and artists. These projects seek to explore similar questions and topics to more traditional scholarly writing, but through different materials and processes.

Most recently, moving Laura's work from autoethnographic analytic writing described in the "Cyborg Bodies" chapter of this book to art and design practice-based research, she has codesigned, with queer crip fashion designer Sky Cubacub of Rebirth Garments, a bathing suit that accommodates her insulin pump as a way of navigating the politics of visibility and invisibility as a disabled person with type 1 diabetes. Laura also collaborated with interdisciplinary artist Itziar Barrio on a series of sculptures that use text and data from her "smart" insulin pump. Some of the sculptures, which are made of concrete and spandex, feature alert and alarm data annotated with autoethnographic field notes printed directly on the circuit boards, and other sculptures are data-driven in that the data are used to create subtle movements such as writhing and breathing. Laura interprets these movements as reminiscent of the many nights of sleep deprivation that she endured while using the technology. The sculptures were displayed in Barrio's solo exhibition at Smack Mellon gallery in New York in spring 2023. The exhibition's title—*did not feel low, was sleeping*—was drawn from Laura's field notes. This project developed a different affective relationship with technology, subjectivity, and disability by engaging with themes around labor, care, and failure while, at the same time, exploring diverse understandings of data, authorship, and collaboration.

Conclusion

Through a range of creative works, this chapter has demonstrated the ways in which cyborg culture—whether drawing on the concept of cyborg as content or method, or even becoming cyborg—might be understood and experienced in everyday life, popular culture, electronic music, and avant-garde art. Here we illustrated the ways in which the figure of the cyborg is present in works across a range of genres. These works explore themes including identity, speculation, and imagination, expansion of the senses, experimentation with technology, hybridity and remix, the glitch, and feminist politics of technology. In closing, while cyborgs have traveled widely in popular culture, we argue that they are still often misunderstood creatures, as we will discuss in more detail in the next chapters.

A MANIFESTO FOR CYBORGS

"By the late twentieth century, our time, a mythic time," writes Haraway, "we are all chimeras, theorized and fabricated hybrids of machine and organism—in short, cyborgs. The cyborg is our ontology; it gives us our politics. The cyborg is a condensed image of both imagination and material reality, the two joined centers structuring any possibility of historical transformation."[1] With this powerful declaration that "we" are all cyborgs, and that the cyborg was a figure that could teach us about the technopolitics of the future as well as being in the world as feminists today, Haraway's 1985 essay, "A Cyborg Manifesto: Science, Technology, and Socialist Feminism in the Late Twentieth Century," ushered in cyborg theory in the humanities and social sciences.

Since this text is foundational to thinking with and as cyborgs, this chapter provides a guide to reading the

"Cyborg Manifesto" that contextualizes it in labor, disability, and feminist concerns. The central tenet of this essay—that technology, bodies, and culture are intimately linked in modern societies, and that it is our job to understand how these linkages are changing social institutions and power relations—is an important part of critical cyborg literacy. The "Cyborg Manifesto" is a complex text, so no single look at it could supply a comprehensive guide. Our reading emphasizes the themes we've focused on throughout this book so far: labor, bodies, and feminism. We also introduce some critiques of Haraway's cyborg that we will delve into in more depth in the following chapter, "Troubling Cyborgs." These two chapters aim to help you deepen the critical cyborg literacy you have developed thus far.

Since the 1980s, thinking with and as cyborgs has been taken up in fields as varied as English literature, anthropology, communication, STS, feminist studies, philosophy of technology, design, and art. This multidisciplinary conversation has sprouted new ways of thinking about the nature of technology, progress narratives, and the relationship between technology, human society, and nonhuman organisms. The terms of these debates have been indelibly shaped by Haraway's improbably popular essay. Today, different disciplinary and practitioner perspectives yield different understandings of what arguments, references, and conversations are crucial to take away from the piece.

Why Cyborgs?

One major question deserves answering before delving into these more specific concerns: Why cyborgs? Why did Haraway see these icons of science fiction as such an apt metaphor for personhood and politics in a socialist-feminist frame in the 1980s? And why has cyborg theory persisted to 2023, when we are putting the finishing touches on this book?

In many ways, using the cyborg as a figure in feminist theory is an improbable move. Cyborgs were an invention of the US Cold War era military-industrial complex. They were dreamed up as tools for war and imperialism. As discussed in chapter 1, the term *cyborg* was first published in a 1960 paper in *Astronautics*, following the introduction of the term at a military conference by Manfred Clynes and Nathan S. Kline.[2] Linguistically, *cyborg* was a portmanteau of cybernetic organism. Practically speaking, cyborgs were an invention of military-adjacent cybernetics researchers aiming to solve the problems posed by space flight and other situations that put the human body and psyche under extreme stress.

In the 1960s, cybernetics had only recently become an identifiable scientific field in the United States, United Kingdom, and USSR. This blossoming area of inquiry was defined as the science of communication and control by Norbert Wiener in his 1950 book, *The Human Use of*

Human Beings.[3] For Clynes and Kline, the communication and control system that mattered for the creation of future cyborgs in space flight was communication between animal biology and engineered machine systems. Cyborgs would use chemicals like hormones, electric signals, and direct mechanical manipulation to transmit "messages" between the mechanical components and biological systems to help preserve and sustain life in future space exploration missions, for example. The ideal cybernetic organism, whether human or animal, would unite these elements together into an "exogenously extended organizational complex functioning as an integrated homeostatic system unconsciously."[4]

The automatic, seamless, unthinking integration of machine and biology was key to real-world cyborgs in their original conception. Making cyborgs a reality was a problem to be solved through the combined efforts of mathematical modeling, physiology, and mechanical and electrical engineering. Successful engineering would enable military and economic domination of the heavens. While many early cybernetics researchers had strong political views, no one at the time could have guessed that the cyborg would be taken up as a feminist icon two decades later.

Haraway's essay acknowledges these military origins of the cyborg, yet seeks to recuperate it for feminist ends. Haraway deploys the cyborg as a figure for dismantling the binaries and boundaries constructed between men and women, "First World" and "Third World," animal and human, and human and machine. Cyborgs, repurposed for liberation, represent an "ironic political myth . . . faithful as blasphemy is faithful" to their militaristic origins. Haraway rewrote their story to make them relevant to the traditions of socialism, feminism, and materialism that she cared deeply about. As a "creature of social reality as well as a creature of fiction"—fictional in both the sense that they were rapidly taken up in pop culture and that twentieth-century engineers never delivered on the promise of suit-free space flight—the cyborg also opens up conceptual space for imagining how to appropriate other technologies of communication and control to aid in local and global struggles for the liberation of the oppressed.[5]

Haraway's own frame of reference for liberation struggles came from her participation in feminist movements in the United States in the 1970s and 1980s. The cyborg presented both dangers and opportunities for feminist thought. The cybernetic logic of communication and control that inspired cyborgs also exerted control over women in workplaces and private homes worldwide, from writing to advanced telecommunications, and from the housewives of Silicon Valley to the women and girls assembling home computers in 1980s' Singapore and Taiwan. At the same time, the cyborg offered a way out of salient debates in the 1980s about whether nature or nurture was to blame for the persistence of patriarchy in the face of feminist agitation and technological advancement. If feminism did not birth cyborgs, by the 1980s it could certainly use them to imagine new ways to shift the tides of global high-tech capitalism and improve the lives of women.

In feminist science studies, cyborgs symbolize the bridging of three domains of social activity, each critical to understanding how work, bodies, and feminism anchor politics and experience: the "territories of production, reproduction, and imagination."[6] Cyborgs open up new ways to think about capitalist modes of production, particularly the emergent forms afforded by the global outsourcing, shipping, and communication regimes that were already evident in the 1980s. Cyborgs prompt reflection on reproduction, a core feminist concern and political issue for all bodies that gestate and care for others across the generations. A cyborg lens treats human bodies and bodily processes as part biological and part crafted at the same time. It can therefore disrupt assumptions that nature alone determines how bodies act and are experienced in terms of gender and reproduction as well as with other bodily traits and capacities. Finally, cyborgs are a way to imagine alternative possible futures. They have been figures of future potential progress in both scientific theory and science fiction since the term was coined six decades ago. While imagination does not replace the political and material work of feminist and other activisms, it is a prerequisite for proposing the more just and inclusive futures that activists in all domains pursue in the world.

Work

The same cybernetic communications technologies that made it possible to dream of cyborgs in the 1960s have reshaped how most people around the world work today. In our earlier chapter about work, we talked about some real-life examples, such as the ways that Amazon tracks workers and automates different parts of its shipping operations. Fictional cyborgs, like the workers in *Sleep Dealer*, offer cautionary tales about the extension of surveillance techniques over workers in their personal and professional lives. Here in the real world, communications technologies have remade work in much more mundane ways. These reconfigurations are both less obviously threatening and more pervasive. Critical cyborg literacy allows an observer to apprehend the high-tech and capital-intensive networks that sustain gendered and racialized patterns of labor and strategic wealth inequality in the modern world of work.

In the "Cyborg Manifesto," Haraway highlights the networks that drove the production and use of microprocessors and high-powered computers in the 1980s. While these technologies might seem miraculous on their face, as though they are immaterial or "made of sunshine," they are, like the people who make them, entirely material.[7] Women are intrinsic to these networks. In fact, as historians like Mar Hicks and Jennifer Light have documented, women built and operated many early computers, like the ENIAC computer designed at the end of World War II using military funding, despite the later shift to a masculine coding of computing work.[8] They were attractive employees in part because they could be paid less than men would accept for the same work. There was a growing pool of highly educated women available to do the work too. Components like microprocessors have often been made by the hands of women and girls as well, in factories in Singapore in the 1980s when Haraway wrote the "Cyborg Manifesto" and today in Shenzhen, China. The targeted recruitment of women in computing echoes the nineteenth-century history of industrial textile production in Manchester, England, and Lowell, Massachusetts. In these historic instances, young women and girls were also intentionally recruited to work for low wages under dangerous conditions to propel one of the most important, high-tech areas of global trade.

Through such opportunistic targeting of women and girls, past and present, Haraway argues that "women's historical locations in advanced industrial societies . . . have been restructured partly through the social relations of science and technology." New technologies, like the loom in the nineteenth century and programmable computer in the twentieth, have created new forms of work. The reserve labor force of women has repeatedly been recruited to do these new types of work at a discount. A cyborg perspective, Haraway suggests, can help an observer navigate the frequently contradictory web of claims about immaterial essences and material realities that surrounds women's high-tech work. "The 'New Industrial Revolution'" of the 1980s, Haraway writes, "is producing a new worldwide working class. The extreme mobility of capital and the emerging international division of labor are intertwined with the emergence of new collectivities, and the weakening of familiar groupings."[9] Technology work represents a paradigm of deepening divides between the haves and have-nots at a global scale. People remain fixed, held in place by family, tradition, and political borders, and old forms of solidarity like trade unions can no longer protect workers from exploitation. Meanwhile, capital and technology freely circulate, optimizing for buying low and selling high in global labor and consumer markets.

The remaking of work is not only about women, however. Haraway contends that there has been a broad-based reconfiguration of work in the global economy. Borrowing from political theorist Richard E. Gordon, she refers to it as "the homework economy." The *homework economy* denotes "a restructuring of work that broadly has the characteristics formerly ascribed to female jobs, jobs literally done only by women." Haraway goes on to describe the homework economy's effects:

> Work is being redefined as both literally female and feminized, whether performed by men or women. To be feminized means to be made extremely vulnerable; able to be disassembled, reassembled, exploited as a reserve labor force; seen less as workers than as servers; subject to time arrangements on and off the paid job that make a mockery of a limited workday; leading an existence that always borders on being obscene, out of place, and reducible to sex.[10]

In the emergent homework economy in the 1980s, many of the trends toward labor precarity seen today in the global economy were already becoming evident. The Fordist family wage, a wage earned by a male head of household that was high enough to support an entire nuclear family, was eroding. The administrative, service, and care work jobs open to women were increasing, but few paid enough to support a single adult, let alone an entire family, when working a reasonable number of hours per week. As middle- and working-class women took on more responsibilities to provide wages for their families, they were still responsible for work inside the home, like childcare and home maintenance, creating double and even triple shifts—truly a mockery of a limited workday.

Women's roles continue to be seen as expendable and thus are underpaid, while work coded masculine increasingly becomes expendable. As a result, men are losing permanent, well-paid employment, and their families lose the social status and stability it once provided. This is how work is becoming "feminized" across the board. Such trends are especially evident for care-focused workers like home health attendants and special education teachers. Both men and women work for wages that are often at or below the poverty level, even with postsecondary degrees. And as those jobs that used to be seen as women's work change from nice-to-have to absolutely necessary for the health of national economies, there are fewer options to quit or change occupations for better pay when confronted with discrimination in the workplace, or difficulties balancing work and family life. Meanwhile, in the home, there is an "ensuing intensification of demands on women to sustain daily life for themselves as well as for men, children, and old people" due to "the collapsing welfare state."[11]

Haraway's discussion of work in the "Cyborg Manifesto" should serve as a warning that the benefits of distributed work are likely a mirage. Instead, critical cyborg literacy encourages a view of work that considers how it operates in historically and geographically variegated ways. New workplace arrangements previously hailed as revolutionary have further entrenched class, geographic, racial, and gender divides rather than overcome them. Further, critical cyborg literacy calls on us to consider the messy integrations of seemingly different domains: work and home, identity and industrial production, self-determination and integration into a rigid system. Working flexibly might just mean working more.

When it comes to work, a cyborg perspective doesn't offer easy answers. In fact, it offers as many disappointments and cautionary tales as causes for hope. Critical cyborg literacy keeps bodily reality and transcendent aspirations in the same frame to strategize in realistic ways to maintain dignity and bodily autonomy, no matter what comes next.

Bodies

In Haraway's reading, the cyborg is a figure that helps us imagine not only new social arrangements around work but also new understandings of human bodies. Rather than getting mired in debates about the role of nature versus nurture in shaping our bodies, cyborgs—"simultaneously animal and machine, who populate worlds ambiguously natural and crafted"—supply imaginative possibilities for undoing dichotomies that are too often used to classify and control people. In the opening pages of the "Cyborg Manifesto," Haraway asserts,

> The cyborg is a creature in a post-gender world; it has no truck with bisexuality, pre-Oedipal symbiosis, unalienated labor, or other seductions to organic wholeness through a final appropriation of all the powers of the parts into a higher unity. . . . The cyborg skips the step of original unity, of identification with nature in the Western sense. This is its illegitimate promise that might lead to subversion of its teleology as Star Wars.[12]

The figure of the cyborg, with its roots in both biological nature and human-made technology, helps readers confront the hybrid character of the human body. Gender is a primary site of this disruption. The cyborg figure's complicated "birth" in a laboratory lets it bypass the nature versus nurture debate entirely. Taken to its logical end point, cyborg politics throws into doubt *all* claims to pure and simple explanations of the relationship between gendered or sexed bodies and power. The cyborg's ironic play with sex, gender, and reproduction provides feminist analysts in particular a familiar foothold into broader questions about the "nature" of the human body.

The manifesto is concerned with breaking down more boundaries than just gender. Haraway also argues for the breakdown of the boundaries between human and animal, organism and machine, and physical and nonphysical. Each of these boundary breakdowns has implications for the human body. Environmental movements demonstrate how "the boundary between human and animal is thoroughly breached." Our fates as living beings on planet Earth, for example, are dependent on everyone's ability to breathe clean air and consume clean water. The "leaky distinction" between "animal-human" organisms and machines has "made thoroughly ambiguous the difference between natural and artificial, mind and body, self-developing and externally designed, and many other distinctions that used to apply to organisms and machines." And the (apparent) blurring of the physical and nonphysical by electronic computers recalls older debates about the power of language and writing. "People are nowhere near so fluid" as text or computer code, "being both material and opaque."[13] Cyborgs make us face the constraints of the flesh while also promising us a future in which the body will be profoundly changeable and fluid.

For some readers of Haraway's work, such critiques of the categories of human, animal, machine, and the physical overshadow the reality of living as a cyborg—as a body both crafted and born, and hence simultaneously powerful, political, and vulnerable. Some of the most trenchant critiques (and in our view, most generative extensions) of the cyborg's disruption of the human body come from disability scholars and activists. Chief among Haraway's critics is Jillian Weise, whom we discussed in the earlier chapter on cyborg bodies. In a 2018 essay, "Common Cyborg," Weise writes, "The manifesto coopts cyborg identity while eliminating reference to disabled people on which the notion of the cyborg is premised. Disabled people who use tech to live are cyborgs. Our lives are not metaphors."[14] As a corrective to Haraway's incitement to future-oriented action and metaphoric abstractions, Weise outlines the violence done to disabled people in the present: the compulsion to accept unwanted technological interventions, increased vulnerability to bureaucratic and physical violence, and unpredictability of living in a body that is integrated with technology. Real cyborgs, as Weise defines them, suffer in today's world, which too frequently values technical quick fixes over the acceptance of disabled life.

These challenges to and extensions of Haraway's figure of the cyborg expand the body politics of cyborgs. Critical cyborg literacy challenges gender and sex binaries, and by extension throws biologically determinist claims that nature trumps nurture into question. It also calls attention to the macrolevel economic and care infrastructures that shape disabled life—dynamics that we detailed earlier in this book. Entwined in these axes of identity and power that shape bodily well-being are race and colonialism as well. Conditions of work, activism, and care combine to shape the bodily potentialities of lived experience in the cyborg world of today. Technology might accelerate the effects and changes of these systems, but flesh and selfhood are relevant determinants as well.

Feminism

While the "Cyborg Manifesto" has come to have a wide and interdisciplinary audience, originally one of the most direct areas of intervention was feminist theory in the 1980s. In her sketch of the landscape of feminist debates at the time, Haraway maps out positions for Marxist/socialist feminism, radical feminism, and feminist psychoanalysis. She observes that each ultimately relies on an essentialism of the experience of sexed bodies along with universalizing claims to speak for all women across time and space. Debates about the relationship between white, middle-class, US identity as well as silencing or oppression within feminisms, she notes, had recently started to destabilize such grand theoretical gestures. "None of 'us,'" Haraway writes, "has any longer the symbolic or material capability of dictating the shape of reality to any of 'them.' . . . White women, including socialist-feminists, discovered (that is, were forced kicking and screaming to notice) the noninnocence of the category 'woman.'" And yet given the retrenchment toward the patriarchal family, digitally enabled colonization of non-Western people and resources, and steady erosion of labor protections in the United States where she was herself situated, Haraway worried that there remained a need for "political unity to confront effectively the dominations of 'race,' 'gender,' 'sexuality,' and 'class.'"[15]

The cyborg offered a path out of debates about competing ways to essentialize the category of "woman." As creatures both natural and crafted, the cyborg could speak to the ways that sex and gender are experienced in the body as something both familiar and imposed. Haraway positions the cyborg as a response to the essentializing currents in feminist thought at the time of her writing. She maintains,

> With the hard-won recognition of their social and historical constitution, gender, race, and class cannot provide the basis for belief in "essential" unity. There is nothing about being "female" that naturally binds women. There is not even such a state as "being" female, itself a highly complex category constructed in contested sexual scientific discourses and other social practices.[16]

Crucially, cyborg consciousness—or what we have called critical cyborg literacy—emphasizes tools for liberation that are not based on deterministic ideas about biological sex, the primacy of labor over all other social activities, or a commitment to concepts of identity unmoored from real bodies and histories. Because cyborg bodies are both/and—animal and human, organism and machine, physical and nonphysical—they can encompass apparently contradictory or frictional understandings of what identity is, and how it operates in the world. The cyborg is a capacious figure for feminists who aim to root identity in networks of active affiliation rather than in the assignment or adoption of immutable, binary qualities.

The feminist interventions made by the "Cyborg Manifesto" constitute, perhaps, the essay's most fraught legacy. In particular, Haraway's own positioning as white, middle class, highly educated, and affiliated with prestigious institutions raised questions starting quite soon after the manifesto's publication in 1985. As recently as 2020, Haraway's whiteness has motivated suspicion of the manifesto's engagement with Chicana and other (to use the term of the time) "Third World" feminisms.[17] We discuss some of these concerns at more length in the "Troubling Cyborgs" chapter later in this book.

Despite critiques, the "Cyborg Manifesto" persists in the feminist theory canon because it expands the purview of feminist theory into the masculine-coded territory of technology. Elsewhere in second-wave feminist studies, scientific assumptions about gender and sex were becoming a valid topic of feminist critique. Anthropologists like Sherry Ortner were working to delink the concepts of gender mobilized in ethnographic research from Western biological assumptions about sex, while other travelers in Haraway's cohort of feminist science studies scholars, such as Sandra Harding, Evelyn Fox Keller, and Emily Martin, were opening up scientific cultures and practices to feminist analysis.[18] The "Cyborg Manifesto" makes clear that technology, too, was ripe for feminist investigation. From a feminist perspective, technology, and particularly emerging computing technologies, could not be considered neutral or self-propelling. Rather, technology had to be confronted as an invention of human societies that was called into service to operate as tools for those who controlled it within existing systems of power: patriarchy as well as racism, class oppression, and colonialism. Because it was made by humans, it was also changeable and political. Hence it established technology as an appropriate topic of study and intervention for feminists, despite any cultural ideas to the contrary.

Ironically, the imperfect and messy human origins of advanced technologies in the late twentieth century are exactly what made cyborgs promising targets for feminist appropriation and recuperation. Cyborgs, rooted in militarism along with colonizing impulses aimed outward into the stars, have not yet realized their full potential outside science fiction. Yet because their very bodies blur

The "Cyborg Manifesto" persists in the feminist theory canon because it expands the purview of feminist theory into the masculine-coded territory of technology.

the boundary between human-made and naturally reproduced, the hopes of abundance, reduced suffering, and
easy adaptability that they encode can still serve as an inspiration for feminists while putting to rest fractious debates about nature versus nurture. The cyborg future laid
out in the manifesto is not free of struggle, but it is more
forthright about the hybrid character of humanity—that
we are both biologically and culturally shaped—and the
inherent risks in capitalist ambitions to advance and progress. The figure of the cyborg offers an enduring provocation for feminists to critically engage with the high-tech,
digitized world that surrounds us.

Conclusion

"From one perspective," Haraway notes, "a cyborg world
is about the final imposition of a grid of control on the
planet. . . . From another perspective, a cyborg world
might be about lived social and bodily realities in which
people are not afraid of their joint kinship with animals
and machines, not afraid of permanently partial identities
and contradictory standpoints. The political struggle is to
see from both perspectives at once."[19] The cyborg's legacy
as an avatar of political struggle is one of the things that
motivates both of us to return to it again and again as an
analytic tool. The clarity with which critical cyborg literacy

The figure of the cyborg offers an enduring provocation for feminists to critically engage with the high-tech, digitized world that surrounds us.

can reveal the politics of technology is one of the most important legacies of the "Cyborg Manifesto."

Technology does not simplify social life on its own. Instead, it extends and expands the legacies of struggles against domination as well as the constant innovation of new forms of bureaucratic control over workers, women, and colonized people. The complications introduced by novel technologies can bring these ongoing struggles into sharp relief. Often, the immediate harms of new methods of production, reproduction, and imagination are curbed through regulation.

But critical cyborg literacy calls us to greater action. It asks us to fundamentally rethink and dismantle the very structures of power that technologies support. Further, a cyborg perspective trains us to recognize human-designed systems of oppression as the enemy of freedom and flourishing. In other words, the material artifacts we call "technology" are not oppressive on their own, and many can be creatively repurposed as tools of liberation. Cyborgs are emblematic of this, and cyborg theory affords us conceptual tools to fight oppression in our own lives and work.

TROUBLING CYBORGS

So far, we have shown how critical cyborg literacy can enhance our understanding of real-world domains of activity,
including work, health, reproduction, and creativity. The
previous chapter provided more conceptual background, dis-
cussing how the concept of the cyborg has developed from
its origins in engineering to its present significance in social
theory. In this chapter, we want to engage more deeply with
scholarly and activist critiques of Haraway's figure of the
cyborg to illustrate how cyborg theory continues to evolve
in conversation with other important concerns. Heeding
Haraway's own call to "stay with the trouble," this chapter
turns to the trouble with the figure of the cyborg. Specifically, we consider challenges to the whiteness of the cyborg
and critiques from disability scholars. We feel these critiques
are crucial, and listening to them can improve critical cyborg
literacy both for us and you, the readers of this book.

In the world of innovation and concrete applications, cyborgs also cannot escape their own histories, which are born of militarism, colonialism, and control. Further, the origins of cyborg theory are situated within particular traditions of white, ableist feminism. Even some of the most visible cyborg futures are complicit with the powerful. For example, cyborgs are often conflated with the techno-optimistic transhumanism—a belief that humans eventually will leave their bodies to merge seamlessly with computers to achieve enhanced intelligence and ultimately immortality—of Silicon Valley billionaires like Elon Musk and Mark Zuckerberg. We argue that these visions of cyborgs are misunderstandings of their true revolutionary potential and seek to trouble these visions in this chapter too.

Cyborgs are not innocent, but critical cyborg literacy can be a tool for liberation when carefully developed. Feminist scholar Alison Kafer, for instance, points out that on publication of the "Cyborg Manifesto," some readers interpreted it as antifeminist, uncritical, and technoutopian.[1] Yet time has integrated this document into the feminist and critical technologist canons because it does not shy away from making the histories, presents, and futures of technology accessible to critique and reconfiguration. Critical cyborg literacy emphasizes the critical and political potential of the cyborg.

For Haraway, "staying with the trouble" suggests an ability not to rush toward a utopian future in which

Cyborgs are not innocent, but critical cyborg literacy can be a tool for liberation when carefully developed.

problems do not exist but rather learn "to be truly present." She writes, "The task is to make kin in lines of inventive connection as a practice of learning to live and die well with each other in a thick present. Our task is to make trouble, to stir up potent response to devastating events, as well as to settle troubled waters and rebuild quiet places."[2] In this chapter, we sort out the trouble with cyborgs with the help of scholarship that both critiques and engages with cyborg theory from a range of perspectives, including gender, race, and disability studies. This chapter departs slightly from the rest of the book by engaging more with complex scholarly ideas in disability studies, Black studies, critical race theory, and anthropology. For scholars in these areas, cyborgs cause trouble.

Cyborg Trouble in the Real World

To set the stage for these conceptual critiques and extensions of critical cyborg literacy, we first present two real-world examples that we think represent cyborgs gone wrong: Musk's Neuralink brain-machine interfaces and Zuckerberg's virtual reality world, the Metaverse. Cyborgs and cybernetic principles have become deeply embedded in the imaginaries of Silicon Valley founders. Cybernetic dreams of hybrid human-machine systems have birthed the transhumanist movement, which Musk in particular seeks to advance with several of his business ventures. Their reasons for engineering cyborgs, however, tend to rely on tropes of "fixing" disabled bodies, expanding the workday and office into formerly private spaces, seeking to extract capital from those least able to afford it, and privileging the authority of experts over the needs of users. These are not the goals that we imagine for cyborgs, nor are they the kinds of cyborg futures that we hope readers will build with the tools of critical cyborg literacy at hand. Instead, the cyborgs that Musk and Zuckerberg offer the public extend capitalist, racist, ableist, and sexist logics. They simply maintain the status quo behind a cloak of novel technological gadgets.

Musk's company, Neuralink, founded in 2016, claims that it will be able to implant computer chips into the human brain in order to create a brain-computer interface. According to one company promotional video, this technology will "solve important brain & spine problems with a seamlessly implanted device."[3] The video uses three live pigs—Joyce, Dorothy, and Gertrude—to illustrate that it is possible to insert computer chips into the brain, read the brain waves as data, and externalize it as a visualization and audio beeps. The demo then shows Gertrude walking on a treadmill. The narrator alleges that the device predicts the position of the pig's muscles based on brain waves.

The device is the size of a large coin and shares the rose-gold aesthetics of the latest generation of mobile technologies. Throughout the demo, Musk goes to great lengths to convince journalists that the implant process will be invisible, seamless, and "look totally normal" once the device is installed. At one point, he compares it to a "Fitbit in your skull with tiny wires." Musk adds that it will be easy to insert and remove via an outpatient procedure performed by a surgical robot, without any bleeding or neural damage. Finally, Musk compares this technology to other widely available and inexpensive medical procedures such as Lasik. The goal, he suggests, is that we may one day be able to save and replay memories, store them as backups, and download them into a new human or robot body. Such claims echo the belief in transhumanism and recall the science fiction program *Black Mirror*'s "The Entire History of You" episode in which a "grain" memory implant can record everything that we do.

In the case of the Metaverse, Zuckerberg offers a vision for integrating virtual reality, augmented reality, and video in order to create real-time interactions in the digital world. In the official company video announcing Facebook's rebrand as Meta, Zuckerberg already looks very much like a digital Lego avatar come to life with his customary closely cut hair. He begins by claiming that this continues the company's commitment to "bringing people together" and connection. From there, he goes on to introduce a suite of products under the umbrella of Horizon, including Horizon Home and Horizon Worlds. Far from supplying a revolutionary product, Zuckerberg instead appears to be selling us a version of the Zoom life that many grew accustomed to during the COVID-19 pandemic. Passing references to ongoing concerns about Meta's record on privacy, safety, consumer choice, creativity, and environmental impact are added in to quell fresh rounds of criticism of the company.

The innovation Zuckerberg holds out ultimately seems rather minor: the addition of three-dimensional digital avatars appears like an incremental improvement over the ubiquitous black boxes of the Zoom form factor. This intervention adds a thin layer of virtual cyborg embodiment onto existing modes of life, including arrangements of work, wealth, and bodies. It seems probable that the true result of Horizon platforms will be the transformation of more of the physical world into digital assets—clothes, images, artwork, voices, and gestures. Yet even as he claims these minor improvements to virtual communication represent a kind of technological revolution, Zuckerberg asserts that the company does not "want to lose too much money on this program overall." The promise of revolution is tempered in the same breath by the exigencies of cost accounting. Even in the pitch, the potential of Zuckerberg's cyborg offering is hamstrung by the capitalist system in which it has been developed and deployed.

It is relatively easy to throw punches at the latest examples of Silicon Valley techno-optimism. So often, tech moguls promise linear progress toward a transhuman-ist vision of a better version of universal humanity that can be achieved through technology. But it is important to carefully break down the rationale for these supposed innovations. Critical cyborg literacy helps us truly under-stand what is at stake.

The claims of tech evangelists like Musk and Zucker-berg are problematic for a number of reasons. We will ex-plore them in this chapter. First of all, they take as obvious the category of the human. They expect that their tech-nologies will improve life for all humanity, even though they are one-size-fits-all and expensive to use. Most likely, the humans who will be improved using their technologies will be white, male, able-bodied, and economically well-off. They do not engage meaningfully with the forms of difference—race, class, gender, disability, nationality, and more—that users may bring to their experience of the technologies. They offer trite comments about economic cost, access, and "diversity and inclusion," but ultimately trust the invisible hand of the market to deliver "trickle-down technology." That is, they expect costs to start high and then go down as the product is more widely distributed.

In addition, the innovations offered by Neuralink and Meta bolster technodeterministic solutions to "the prob-lem of being human"—our physical bodies and disabilities. When they promise that these tools will improve our lives, they conflate the human with the machine in an overly simplistic way. Their technologies offer quick, in-home, or outpatient solutions to human desires for long, healthy lives and interpersonal connection. Tools like Neuralink normalize the human as able-bodied; disabled people with certain physical or mental disabilities will be "fixed" through technology. Beyond their pitches, entrepreneurs like Musk and Zuckerberg suck all the air out of the room when it comes to framing the potential of these technologies. They control the rhetoric and troll their critics using a wide variety of tactics, including lobbyists, robust legal teams, and Twitter mobs.

Finally, while promoting their own benevolence, powerful techno-optimists understate and wave away the problems that technologies introduce, such as heightened threats to privacy, security, and interoperability. For instance, a recent article about bionic eyes becoming obsolete provided a glaring example. When Second Sight, a company that makes retinal implants, decided to stop making them due to bankruptcy, the worlds of hundreds of blind people using the technology abruptly went dark with no support or plans to transition them to other medical devices.[4] In this case, exuberant claims about technology and a linear narrative of inevitably increasing human-technology fusion allowed a company to burn brightly for a moment. In the end, though, disabled people were harmed when the economic winds began to blow in a different direction. From a critical cyborg literacy stance, this is a common risk of transhumanism. Companies and founders profit by putting data-collecting sensors in our brains, bodies, and physical spaces, and leave their users high and dry when their time is up.

The Problem with Cyborg Theory

The debates over whether cyborg theory is truly liberatory or ultimately conservative have motivated us to trouble cyborgs in this chapter. At the highest level, the trouble with cyborgs is their origin story. Cyborgs are born of (and center) Western dualisms, despite their specific purpose in critiquing those notions. This problem is not only common to cyborgs. A range of concepts from the Anthropocene to the posthuman to more-than-human and especially transhuman can be accused of such a history. In responding to Western, European notions of the human that coevolved with science, colonialism, and modernity, these concepts set their gaze on a white, male, wealthy, and able-bodied version of the human when they could be looking elsewhere. Moreover, the speed with which concepts like the cyborg have been taken up in both academic and mainstream discourses serves to further universalize the meanings of these terms. As such, the histories, contexts, and situatedness that are essential for the critique are removed and forgotten.

Cyborgs cause trouble precisely because of their popularity. They capture our attention in movies, science laboratories, trade shows, and museum exhibitions as well as in mainstream media stories and on social media platforms as marketing influencers. Their frequent appearance in popular culture naturalizes them. As we've illustrated in this book, cyborgs can be found in controversies around the future of work and cutting-edge cultural practices, and even by looking at our own bodies, health, and medicine. Cyborgs come to represent everything and, at the same time, nothing at all. This chapter explores reasons to reject the cyborg's entry into circulation, its limitations, and lessons for looking elsewhere and otherwise.

Black Cyborgs / White Cyborgs

The question of race causes trouble for cyborgs. In the television series *Lovecraft Country*, one character, Ruby Baptiste, shifts from Black to white with the use of a potion. On the one hand, this shape-shifting character seems to illustrate the blurred boundaries around racial categories. Her transformation hints at an apolitical, postracial possibility, yet at the same time clearly portrays the fact that race is primarily a social construct. On the other hand, Ruby's character arc further reifies these boundaries by depicting the ways in which the Black character's mobility is limited in the context of the segregated 1950s' United States. (It should be noted that the book on which the series is based was written by a white, male author.)

This example demonstrates the importance of the question about whether and how deeply considerations of race have been embedded in the cyborg. Like Ruby's character, the figure of the cyborg fits comfortably with a liberal postracial political imaginary that ignores Black people's material experiences of structural inequality and oppression on the basis of skin color. Critical cyborg literacy must consider how race, racism, and colonization shape the liberatory potential of cyborg technologies, and redirect thoughts and practices that are founded on exclusion and oppression.

The first broad critique of the cyborg that we will outline extends reflections from the previous decade about who is included in the category "human." Scholars and activists alike argue that it makes little sense to move beyond the category of the human—into new relations with machines, nature, and each other—when so many people are still being dehumanized. If this sounds abstract or extreme, one only need look at the efforts to prevent Black people from voting in the United States, unequal health outcomes during the COVID-19 pandemic for lower-income and disabled people, rates of police violence against and incarceration of Black men, and patterns of underemployment for Black women.

In her "A [White] Cyborg's Manifesto," researcher and academic Julia DeCook contends that Haraway's figure of the cyborg "exacerbates categories of difference" due to the ways in which it is implicated in "Western, patriarchal violence" and that "'cyborgs' are not outside of the politics within which they exist." DeCook draws on her own experience as a "mixed-race, military child"—born of a US soldier and Korean woman—in order to question the ways in which the cyborg centers whiteness around Western epistemologies around identity. Like many others scholars that critique concepts related to posthumanism, DeCook maintains that the cyborg becomes a new kind of technoutopian universal "where technology is seen to be emancipatory, rather than oppressive."[5] While this technological euphoria has been attached to each technological breakthrough for generations and is certainly evident in the past thirty years, recent debates over topics such as AI bias, "smart" surveillance, and working conditions in "platform capitalism," to name just a few, have illustrated a growing and louder technopessimism. Scholars such as Christina Dunbar-Hester, Morgan Ames, Mar Hicks, Safia Noble, Ruha Benjamin, and Virginia Eubanks have shown that usually technology continues to maintain the status quo, and even exacerbate existing structural inequalities around gender, class, race, sexuality, and ability. Rather than conflating the cyborg with naive technodeterminism, we argue that both technoutopianism and technopessimism can be found within the cyborg.

DeCook asks, "Who gets to be a cyborg?" asserting that it is a privileged identity that on the one hand, ignores race, and on the other, draws on the identities of women of color. "The cyborg does not come without baggage," writes DeCook, and indeed the very question of who counts as human is at issue. In pointing out the limitations of the cyborg, DeCook reminds us that "its enduring mythos is the idea that there is a beyond, a space that is both material and virtual, and one where identity is not necessarily dismantled but rather exists in relation to the technological." Ultimately, she concludes that "technology will not save us. Cyborgs will not save us. But there is hope that in challenging these notions, our humanity will."[6] Yet by making these crude distinctions between technology and humanity, this critique runs the risk of further reifying the categories that cyborgs seek to destabilize and unsettle.

In the book *Habeas Viscous*, scholar Alexander Weheliye ties the concept of the human to scientific and technological progress. He positions the version of humanity illuminated by Western science as selective and narrow because inquiries into the "nature" of humanity have often been conducted without explicitly attending to how race shapes the experience of particular humans. Weheliye notes,

> Though the human as a secular entity of scientific and humanistic inquiry has functioned as a

central topos of modernity since the Renaissance, questions of humanity have gained importance in the academy and beyond in the wake of recent technological developments, especially the advent of biotechnology and the proliferation of informational media. These discussions, which in critical discourses in the humanities and social sciences have relied heavily on the concepts of the cyborg and the posthuman, largely do not take into account race as a constitutive category in thinking about the parameters of humanity[7]

A necessary remedy for critical cyborg literacy, following Weheliye, is to specifically address race as an outcome of scientific inquiry and technological progress.

Kalindi Vora and Neda Atanasoski take up this challenge in their analysis of race in cyborg technologies. For instance, they propose that the racial capital regimes into which robots and other animated technologies are becoming seamlessly integrated are rooted in what they call "the surrogate relation."[8] Surrogates do the dirty work of society; they do the care work, wage war, and clean up. Formerly, this work was done by the oppressed classes and races of humanity. With modern technologies, much of this work can be displaced onto sophisticated machines. This preserves racial and class inequalities, however, because the operators of these technologies are still drawn from the least privileged groups in society. Technology seems to offer an escape from drudgery for colonized and racialized people. But in reality, Vora and Atanasoski argue, new technologies simply mask the racialized operations of capital.

Vora and Atanasoski's critique serves as an oblique corrective to the utopian visions of the "Cyborg Manifesto." While technology *may* liberate racialized and colonized workers, including female domestic and electronics workers, this is by no means assured. These workers, especially female ones, may, for example, merely become the caretakers of killing and cleaning machines. Without active intervention, the same groups will still be sworn to drudgery as before. Their labor will continue to enable lives of unimaginable wealth and ease for white people and the wealthy upper classes of the world. Fully realizing the technoutopian promise of cyborg consciousness requires a political vision from the start that is more inclusive of the lived realities of colonized people.

One remedy that these perspectives suggest is a shift toward praxis—a closer relationship between reflection and action. Scholars such as bell hooks and Paulo Freire have argued that lived experience is a form of knowledge about structural inequality and oppression, particularly based on gender and race. For hooks, reflecting on her own childhood, "theory" and "theorizing" offered a respite, a place to make sense of things, a place where she could "imagine possible futures, a place where life could be lived differently." She describes theory as "a healing place." As hooks puts it, "When our lived experience of theorizing is fundamentally linked to processes of self-recovery, of collective liberation, no gap exists between theory and practice."[9] For Freire, writing about awareness and resistance to oppression in Brazil in the late 1960s and early 1970s, praxis was about "reflection and action upon the world in order to transform it."[10] For both of these influential thinkers, theory is not understood as abstract or disconnected from everyday life but rather tightly coupled together.

While the express purpose of the cyborg is to overcome such binary distinctions as that between theory and practice, cyborg theory is frequently presented primarily as a theoretical intervention. We hold that our theory and action must be closely linked. In fact, this has been one of the central aims of this book: to introduce new audiences to the value of thinking like a cyborg to understand the ways in which technology is embedded in multiple realms of everyday life such as work, the human body, and popular culture. By using different lenses to think about the world such as those offered by cyborg theory, different worlds are made possible.

A second way forward, drawing on the Black speculative arts movement and Afrofuturist tradition, discussed above in the "Cyborg Culture" chapter, is based on the potential for expansive creativity, remix, and liberation through technology. Art, media, and music in the Afrofuturist tradition complement the cyborg association with imagination, speculation, and experimentation. The movie *Black Panther*, while both celebrated and critiqued, is perhaps one of the most broadly distributed Afrofuturist works. It illustrates simultaneous attunement to aesthetic and cultural traditions and advancement through the considered use of technology.

A third response, or resolution of these tensions, looks at the ways in which technology and Blackness are connected, such as assertions by scholars and creatives that race itself is a technology.[11] In *Black Futures*, speaking about Black social dance, scholar Jasmine Johnson writes, "The Black body is a technology: it knows, remembers, and utters."[12] Taking in these more capacious understandings of technology—beyond machines and digital technologies—that seem quite consistent with Haraway's broad-based view of the sociotechnical may be another way to carry cyborgs into more critical conversations about humanity and race.

While the figure of the cyborg has been critiqued for its whiteness, Black studies and critical race theory offer some ways for critical cyborg literacy to provide better paths forward. To create technologies for work, medicine, and pleasure that are truly inclusive, these criticisms must be integrated into contemporary cyborg and futurist thought. But race is not the only dimension of the cyborg that has been critiqued in recent years. As we will see in the next section, disability scholars also have complaints about the original framing of the cyborg. They have extended cyborg thinking into new domains, even claiming it as a political figure for disabled people in the new millennium.

Disabled Cyborgs

The question of disability causes trouble for cyborgs. Kafer traces conversations about cyborgs and disability to Haraway's original reference to people whose bodies are hybridized through their use of prosthetics and machines. She shows how Haraway's points were then taken up in both feminist theory and disability studies. Kafer argues that the cyborg figure that traveled into disability studies was apolitical while the cyborg that was reproduced in feminist theory lacked any meaningful understanding of disability. She tries to reunite these lineages in her "close crip reading of the cyborg."[13]

In this reunion, Kafer affirms the potential of cyborgs for opening up "crip futurities." Both the feminist and disability versions of the cyborg distrust essentialist identities, reject notions of wholeness, and encourage coalition politics built on mutual aid and "political affinity rather than biological identity." For example, Kafer observes that "a cyborg politics would not require an amputee, a blind person, and a psychiatric survivor to present their identities and experiences as the same, but rather would encourage the formation of flexible coalitions to achieve shared goals." Kafer contends that we should not "abandon the cyborg" but instead continue to "struggle with the figure" and reimagine it from a "critical crip position" drawing on intersectional perspectives.[14]

One object of Kafer's critiques is the common conflation of cyborgs and disabled people in mainstream media representations of assistive and adaptive technologies. These representations too often ignore the problems that such technologies cause and the economic cost of using them. Instead, they take a techno-optimist stance that positions technologies as saviors of disabled people. In addition, these portrayals are not tied to the politics of disability justice. Rather than educating people about disability or liberating us from Western dualisms, writes Kafer, mainstream accounts only further reify the binaries of natural/unnatural, human/machine, pure/impure, nondisabled/disabled, and normal/abnormal. Such feel-good stories are a kind of "disability porn" that glamorize the ways in which disabled bodies can be "fixed" through the benefits of technology—a phenomenon that Ashley Shew refers to as "technoableism." *Technoableism* describes a "rhetoric of disability that at once talks about empowering disabled people through technologies while at the same time reinforcing ableist tropes about what body-minds are good to have and who counts as worthy."[15]

But these beliefs are not reserved for the mainstream media. Scientists and engineers adopt them too. In 2021, for instance, MIT announced $24 million in new funding for an interdisciplinary bionics center in order to "move toward a future without disability."[16] Many disability scholars and activists regard such statements as expressing the eugenicist attitude that disabled people should not exist.[17] This is in direct contrast to the disability justice perspective that disabled people's experiences represent expansions of humanity's experience, not deficits.

Kafer ultimately concludes that the cyborg is still a valuable figure to think with because of its multiplicity, contradictions, unpredictability, fluidity, permeability, pervasiveness, and ubiquity—its "history—and present."[18] One opportunity to crip the cyborg is to pay more attention to class, efficiency, and productivity with respect to work—an area where disabled people are acutely made aware of their abilities and inabilities. For example, during the COVID-19 pandemic, there was a rapid shift to remote work for all workers. This prompted widespread outcry by disabled people who had been advocating for decades for accommodations to work remotely. Suddenly, when ablebodied people were at risk, remote work was treated as a legitimate mode of employment. Then employers suddenly required most workers to go back to in-person work later in 2020 or 2021, despite eighteen months of evidence that many jobs could in fact be done remotely. Disabled people were once again denied remote work accommodations once in-person work resumed. This "back-to-normal" approach persists despite the fact that COVID-19 is creating new forms of disability. A crip cyborg politics takes up such obstacles for disabled people and workplace-induced disability as targets for worker and disability movements.

A related discussion surrounds who gets to innovate and who is merely a subject of innovation. Disabled people are typically not the designers of technologies that purport to improve their lives, and designers frequently do not treat their perspectives as valuable during the design process. *Disability dongles* are an illustration of this tendency. This term, invented by Liz Jackson, refers to technologies that are designed as solutions to problems that disabled people do not actually have. Jackson and colleagues Alex Haagaard and Rua Williams offer numerous examples in their essay "Disability Dongle," including augmented reality that enforces normative communication patterns on autistic people, smart footwear for blind people, adaptive deodorant, and stair-climbing wheelchairs. According to the authors,

> Disability Dongles are contemporary fairy tales that appeal to the abled imagination by presenting

a heroic designer-protagonist whose prototype provides a techno-utopian (re)solution to the design problem. Disability Dongle rhetoric instills in students the value of a quick fix over structural change, thus preventing them from seeking out, participating in, and contributing to existing inquiry. By labeling these material-discursive phenomena—the designed artifacts and the discourse through which their meaning is constituted—we work to shift the focus from their misguided concern about our bodies to their under-analyzed intentions and ambitions.[19]

Disability dongles serve mainly as public relations schemes or forms of "access washing." Specifically, writes disability studies scholar Louis Hickman in an invitation to a workshop on the topic, "Similar to greenwashing, *access washing* creates the impression of increasing accessibility or eliminating the barriers that affect people with disabilities while actually displacing the harms onto other marginal communities."[20]

As opposed to erasing disabled people or displacing harms, we turn back to the hopeful directions offered by the cyborg. Kafer concludes that "the cyborg, in other words, can be used to map many futures, not all of them feminist, crip, or queer." An engagement with these critiques of the cyborg from disability studies will ensure that disabled people are seen not as passive receivers of technologies but rather as actively engaged in the making of theory, technology, and expanded notions of humanity, ultimately "refusing the erasure of disability from our presents and futures."[21] Contemporary cyborg theory must include and center disability because disability justice includes significant struggles for power over which technologies humans live with, and how.

Conclusion

This chapter has troubled cyborgs by viewing them through the lens of critical theories of race and disability. While it is not possible to catalog all the critiques of the cyborg, we have attempted to account for some of the most important ones here. We recognize that cyborgs themselves are trouble. For one, they cannot escape their own histories. Cyborg histories are rooted in cybernetics, technoutopian dreams of space travel, and psychological experimentation in mental institutions. Politically neutral cyborgs are popular in the mainstream media and films as well as on television, and as such they are often misunderstood. They are frequently conflated with simplistic, technologically deterministic transhumanist fantasies in which humans are upgraded to the point that we overcome the embodied materiality of biology. In our version of cyborg theory, cyborgs are committed to antiracist, disability justice, and feminist politics. The progress they encode is far from predetermined and would not leave anyone behind.

Critics of cyborg theory describe the ways in which the cyborg erases and universalizes lived human experience, capitalizing on the experience of racialized and disabled women without giving them agency. This casts racialized and disabled women as passive bystanders in their own lives rather than participants in change. They also point out that the cyborg takes an overly enthusiastic attitude toward the possibility of liberation through technology.

While we are in agreement in spirit with most of these critiques, we do not see them as a reason to throw out the figure of the cyborg altogether. Responding to these critiques is how the cyborg endures as a useful figure for navigating contemporary life. We argue that these critical conversations are an excellent way to deepen and enrich our understanding of humanity. Like cyborgs, attunement to how race and disability are present in everyday life can expand what it means to be human. Cyborgs can help us think beyond the ways that the category of the human has been understood for several hundred years as exceptional and exclusive. That is, cyborgs can both challenge and continue to inspire critical and creative thinking about the interrelations between humanity, technology, and politics.

Cyborgs can help us think beyond the ways that the category of the human has been understood for several hundred years as exceptional and exclusive.

In the final chapter, we offer some key principles that emerge from the understanding of cyborgs that we have developed through examples in this book. The cyborg presented in Haraway's manifesto is still enigmatic for many readers. As a feminist intervention into discussions around humanity, technology, and politics, there is great value in revisiting the cyborg for the twenty-first century.

CONCLUSION
Cyborg Futures

Cyborgs, as this book has argued, have a troubled past, lively present, and most likely long future in both real and imaginary worlds. We have approached the cyborg as a figure that grounds a specifically feminist analysis of technology and society. Through discussion of the cyborg's history (in fact and scholarship), modern workplace arrangements, the body and disability, and art and design, we have demonstrated how a cyborg perspective remains a relevant and creative analytic stance. We have met many kinds of real-world cyborgs too: disabled people, workers who rely on modern computing infrastructure, and cyborg representations in art.

In concluding, we offer up ten principles that we think should ground a cyborg perspective. These principles are relevant to designers who may hold the responsibility for making future environments and products in their hands. They are relevant to technologists who are building the

Cyborgs, as this book has argued, have a troubled past, lively present, and most likely long future in both real and imaginary worlds.

digital and perhaps quantum infrastructures for tomorrow. Artists might take inspiration from these principles in situating their work or anticipating how they will connect to audiences. Students and scholars might reflect on these principles when reading cyborg theory or seeking to extend it into new domains in their own scholarship. Of course, cyborgs are multiple, so we do not claim to have the final word on what they can mean. But based on our study and experience, these principles will serve to sustain lively conversation about the cyborg and its legacy in the years to come.

Following these principles, you can find a glossary of some key terms, bibliography of works we cited in this book, and recommendations for additional readings from STS, history, design, Indigenous studies, and more. These resources are our way to invite you into a community of thinkers and practitioners who find cyborgs to be a potent resource for thought and action. Like everything cyborg, these resources are necessarily incomplete, limited by our situated perspectives. We invite you to cross-pollinate our recommendations with things that might already be familiar to you. Together, we can build a better cyborg world.

1   Cyborgs Transgress Boundaries

Cyborgs are beings that transgress boundaries. The transgression of the boundary between organism and machine

We invite you to cross-pollinate our recommendations with things that might already be familiar to you. Together, we can build a better cyborg world.

is the core characteristic of cyborgs in both fiction and reality. The cyborg family tree encompasses actual android robots that are programmed to mimic human emotions, science fiction cyborgs in *Star Trek* with digitally enhanced bodies, and disabled people like Laura living in interdependence with medical devices. As a figure in scholarship, the cyborg symbolizes the transgression of human-made social categories like gender, nationality, race, and disability. In theory, cyborgs can prompt us to reexamine our assumptions about the role that nature plays in shaping society in our high-tech moment. In reality, cyborgs live the contradictions that arise when this boundary is not yet fully dismantled.

2   Cyborgs Are Situated

Cyborgs don't just transgress boundaries in an abstract sense. They do so in ways that are specific to their time, place, and material form. In other words, they are situated in history, space, and body. Key works of cyborg theory, like Haraway's foundational "Cyborg Manifesto," have often been interpreted as advocating for an infinitely flexible subjectivity unmoored from histories of race, colonialism, and bodies. But we interpret cyborgs to be relentlessly grounded in specifics. An android is not the same type of cyborg as a disabled person; a diabetic with an insulin pump does not live with the same intimate infrastructures as a person with a heart pacemaker.

For this reason, we think the cyborg remains an excellent figure for intersectional feminism. As conceived by Black feminist thinkers Kimberlé Crenshaw and Patricia Hill Collins, this form of feminism pays attention to the intersecting types of oppression that shape institutions, communities, and individuals. Identities are not merely additive in this formulation. Rather, historically entrenched social structures, like the criminal justice system or normative family structures, and the expectations that go along with identities such as "woman" or "Black" shape the reality people live in today. The cyborg is an apt figure for this sort of complexity because each cyborg's transgressions are specific to its time and place. Cyborgs remind us as well that other futures are possible—futures in which the oppressive structures we take for granted today are broken down and something closer to utopia has become possible.

3   Cyborgs Live in the Real World

Disabled cyborgs like Ashley Shew and Jillian Weise teach us that disabled people are the original cyborgs. Cyborgs are not firstly—or only—fictional. They are real people too. Real-world cyborgs don't always come with all the futuristic bells and whistles that fiction teaches us to expect. A real cyborg living among us might be someone with a cane, a gestating person, or a diabetic with digitally controlled biostats.

Cyborgs are part of everyday life in the modern world, even though they started their history as beings that would inhabit an imagined space future. They are commonly misunderstood to be about innovation and future human evolution. But cyborgs live in the material world, today. Real-world cyborgs find their kinship with future or fictional characters troubling at times. The struggles around identity, ability, autonomy, and wealth that real-world cyborgs face are frequently glossed over in the fictional renditions. The struggle is real, even for cyborgs.

4 Cyborgs Celebrate the Glitch

Cyborgs are born from the combination of biological and mechanical systems that were not built to work together. So cyborg systems are often glitchy. They break down. They need tweaking. They create unintended outcomes—at times bad, and at other times generative. Rather than denying their imperfections, cyborgs celebrate the glitch. The glitch reveals where the boundaries were supposed to be, until the cyborg transgressed them.

Cyborg glitches are deeply political. They show where the boundaries between male/female, man/woman, able-bodied/disabled, colonized/colonizer, Black/white, and nature/culture are ideological but not actual. They reveal, in other words, where lived experience overflows the boundaries that human-made categories like gender, race, nationality, and disability are designed to keep people within. And frequently, they demonstrate what a life beyond boundaries can look like, glitches and all. This is rarely an easy way to live, so cyborg life is also political in that it is marked by struggle. In transgressing boundaries, cyborgs resist them and underscore that they are not as necessary as those in power would have us think.

Cyborgs take solace in refusal. The glitch is generative.

5   Cyborgs Are Infrastructural

Cyborgs rely on technological networks along with human and nonhuman relations to grow, upgrade, and thrive. Cyborgs cannot simply stand alone as perfectly autonomous individuals. The networks in which they are embedded give them durability. But they also make them vulnerable to changes in the system. This balancing act of vulnerability and durability is a characteristic of all infrastructures.

Cyborgs feel this acutely because they mix organism and machine, fiction and reality, and material and political elements. They are subject to multiple infrastructural systems at any given moment. Cyborg infrastructure often becomes most visible on breakdown.

Cyborg infrastructure is embedded in many domains of life. Cyborgs were born out of the military-industrial complex, and many cyborg technologies still link up with defense-initiated infrastructures like the internet. Cyborgs are embedded in the infrastructures of the modern workplace too: gig work apps, office building codes, and transnational shipping systems. They also have a limb or two in creative fields such as film, art, and design. Cyborgs are distributed across many infrastructural systems. Many infrastructures are cyborg.

6 Cyborgs Take Work

A cyborg's hybrid nature is a work in progress. Becoming cyborg is an ongoing process. Parts wear out. Upgrades are released. New components introduce new glitches that need troubleshooting. These lessons that cyborgs teach us apply even to cyborgs without mechanical components. For digital app workers, joints get sore and injured from riding bicycles on rough city streets. For gestational surrogates, carrying babies to term depletes micronutrients from the gestator's body that may or may not be adequately replaced. For women in the homework economy, the cognitive load of caring for children and pursuing paid work dulls their edge in every pursuit. Even before the point of glitch, becoming cyborg and maintaining cyborg life takes effort.

But work is not the same as labor. Compulsory labor under capitalism—work in the digital, offshored, just-in-time economy that just barely provides for survival—is violent. The work of cyborg self-maintenance, by contrast, can be a source of creativity. Caring for cyborgs can be a source of intimate connections. Doing the work is not the same as working to live. Maintaining cyborgs takes work, but cyborgs strive for a utopia where no one starves if they cannot labor.

7 Cyborgs Care for One Another

Cyborgs care about each other and their noncyborg kin, whether human, machine, animal, vegetable, or mineral. Cyborgs are not impressed with individualistic competition; they thrive on interdependence and mutual aid. They understand that the world is relational, and are also aware that not all relations are the same. Cyborgs seek deeply meaningful relations like kinship and other nonkin interactions in which we shape and are shaped by one another.

The world falls apart when we fail to acknowledge how we rely on one another.

Part of care is attunement. Cyborgs watch and listen to the world around them with their many senses. They might speed up when the world slows down or slow down when the world speeds up, introducing the glitch into systems that do violence to their kin. At other times, they provide stability for relations in need of it. The tension they always feel between durability and vulnerability makes them adaptable caregivers.

8 Cyborgs Speak a New Language

Cyborgs speak in specificities. Cyborgs are multilingual, seeking connections far and wide. They put new names to things in order to destabilize old orders. They do not rejoice in abstraction, universals, rationality, or the infamous but nonexistent God's-eye view. Cyborgs seek to name the hybridity, relationality, and interdependence of their existence in new, interesting, and nuanced ways. This takes them ever closer to the creation of a new language.

At the same time, cyborgs acknowledge that not everything can be communicated in words. Cyborgs do not claim to know everything. Some knowledge is incommensurable. A full account of the world can only be achieved by knitting together multiple accounts of reality.

9 Cyborgs Are Political

Cyborgs are not neutral. They are not "just tools." Cyborgs seek to be *tools for justice*. The cyborg world is one where an equitable existence for all is possible, humans and non-humans alike. This is not to conflate robots with people or suggest that they should have rights. Rather, it is about decentering the human and developing new understandings of our relationship with our creations as well as our resources.

Cyborgs understand that imagination requires politics. Cyborgs reimagine the world, not as a new frontier, blank canvas, or clean slate, but as a conversation with existing histories, politics, and situations.

If cyborgs can imagine it, they can also take responsibility for it. Cyborgs rise to the occasion by designing ethics and politics into their plans. There is no tech ethics without radical politics.

10 Cyborgs Cause Trouble

Cyborgs are troubled figures and also cause trouble. In this book, we took up Haraway's call to stay with the trouble, meaning that we must resist the urge to solve problems. Instead, we present challenging topics, questions, and dilemmas. We engage with complicated questions about humanity, technology, and politics as opposed to seeking future solutions that promise to fix everything and erase everyone. We see free-form speculation about the future as troubling. The future will be full of trouble, just like the present—it just might be a different sort of trouble. The challenge is to reduce the burden on those least able to weather it by holding the powerful to account.

Cyborgs are troubled from the moment of their origin. They were born out of visions about space travel based on testing on patients in mental institutions. They cannot abandon this history because cyborgs are always situated.

Cyborgs are ultimately misunderstood. They are exceptionally popular figures in mainstream media, science fiction, and culture. But in many cases, these examples avoid direct engagement with the kind of cyborg politics that we present in this book. They are sometimes conflated with transhumanism, which insists on a belief that humans will eventually leave their bodies and merge seamlessly with computers in pursuit of a more perfect union with technology. Our account of cyborgs, we hope, troubles such popular figures. Cyborg trouble is rooted in everyday life, real bodies, and urgent politics.

GLOSSARY

Afrofuturism
An artistic and intellectual movement that imagined high-tech futures rooted in African and African American art, music, literature, and other media. Afrofuturism flourished beginning in the 1970s, and overlapped with popular culture trends such as disco and house music in the 1970s and 1980s as well as film in the 2010s.

Artificial intelligence (AI)
The use of computers to do things that were previously done by human cognition such as translation.

Automation
The use of technology in manufacturing or other processes in order to produce goods with less human involvement.

Cripborg
This term, coined by Mallory Kay Nelson, Ashley Shew, and Bethany Stevens, is a portmanteau of *crippled* (a word that is considered a slur when used on its own by nondisabled people) and *cyborg*. Cripborgs are disabled people who embrace both their political identity as disabled people and their necessary uses of technology.

Critical cyborg literacy
The cyborg theory–informed perspective about how people live together with technologies that we develop in this book.

Critical race theory
An intellectual movement originating in legal studies in the United States in the 1980s that examines race and racism as the outcomes of long-term structures of power and oppression.

Cyberfeminism
British philosopher Sadie Plant coined this term in 1994 to describe the ways in which feminists were thinking about the internet and related technologies.

Cybernetics

An area of inquiry that sought to understand and engineer self-regulating systems using advanced communication and control techniques. Cybernetics heavily influenced academic fields as varied as anthropology and sociology, information theory, mechanical and aerospace engineering, and ecology, especially during the 1960s and 1970s.

Cyborg

A hybrid of machine and organism (often human) described by cyberneticians in 1960 as a tool for advancing space flight. Cyborgs quickly became popular in science fiction and other areas of popular culture, especially in the forms of androids (humanoid robots) and augmented humans.

Cyborg theory

An area of cultural and feminist theory rooted in the work of Donna Haraway that uses the cyborg as a figure for philosophical inquiry into the relationships between humans, other species, and technology.

Disability justice

A political perspective on disability that demands a remaking of social institutions, from medicine to education to social services, that prioritizes the needs and inclusion of disabled people.

Disabled cyborg

A concept that technology as well as humans can be understood as disabled, as defined by Laura Forlano in her recent writing. This perspective highlights the ways that the technologies that disabled people use are prone to glitches, friction, and breakdown.

Feminism

A style of inquiry and social movement that uses gender relations as a starting point for understanding and challenging how power, privilege, and oppression operate in societies.

Gestational labor

Sophie Lewis argues that human gestation should be understood as a form of labor. By recognizing gestation as labor, feminist and anticolonial activists gain a new angle for critiquing and breaking down oppressive gender relations.

Glitch

An error, misfire, or malfunction in a technological system. Glitches can accidentally reveal the underlying values of a technology's designers or the broader society in which a technology is embedded. Glitches are usually inconvenient, but sometimes empowering due to what they make transparent.

Infrastructure

Susan Leigh Star defines *infrastructure* as the "invisible work" that sustains technological systems and human societies. Further, political theorist Langdon Winner points out that large-scale technologies—the kinds of things we call infrastructure—encode the values of the people who build them.

Labor

Karl Marx and more recent Marxist theorists see labor as a foundational activity of human societies. A core axis of societal power, in the Marxist tradition, is whether an individual provides labor to another person or entity, or whether they pay for others' labor. At times we use *labor* and *work* interchangeably in this book, depending on whether we are talking about labor as a general relationship or the specific work done by a person or group of people in a particular context, like a particular workplace.

Nature

For cyborg theorists, there are few things that can be described as fully "natural." Most aspects of "nature" on Earth have now been shaped by human hands at one time or another. Cyborg theorists tend to see claims about nature or the natural as an invitation to dig deeper to understand *how* nonhuman elements and human ingenuity work together, rather than cleanly classifying some things as "natural" and others as "artificial." Some scholars even use the term *natureculture* or *naturalcultural* to talk about objects and processes that blend nonhuman "nature" and human agency.

Posthumanism

A branch of continental philosophy that considers the ethics of nonhuman agency. For posthumanists, things such as computers and even trees are considered not merely passive objects but rather active participants in the world. Together, humans and things participate in relationships that make up the world.

Robot
A machine that carries out work that used to be done or continues to be done in some contexts by humans.

Situatedness
Donna Haraway refers to "situated knowledges," the idea that each person's perspective is partial and embedded in their own social, cultural, political, and economic contexts. This is in contrast to claims of universal knowledge and truth, which she calls "the god trick."

Technodeterminism
The idea that technology is a primary driver of social and biological change. This belief has been proven wrong by historians and social scientists whose scholarship examines how technologies develop.

Technology
From the perspective of cyborg theory, technology is not merely a neutral tool; it is shaped by humans, and as such, embeds our ethics, values, politics, and notions of justice. In turn, humans are mutually shaped through our use of technology. This is a sociocultural view of technology rather than one in which technology is the dominant factor in change.

Transhumanism
The idea that the human body will become obsolete once human consciousness is uploaded into machines. Transhumanists seek to extend the human life span and enhance human life by using technology, ultimately seeking immortality. This perspective has been embraced by some prominent Silicon Valley tech founders as well as science fiction films such as *Transcendence* (2014).

NOTES

Chapter 1

1. "Hot Robot at SXSW Says She Wants to Destroy Humans," *Pulse*, CNBC, March 16, 2016, https://youtu.be/W0_DPi0PmF0.

2. "Two Robots Debate the Future of Humanity," Digital Acid, July 14, 2017, https://youtu.be/w1NxcRNW_Qk; "Tonight Showbotics: Jimmy Meets Sophia the Human-Like Robot," *Tonight Show*, April 25, 2017, https://youtu.be/Bg_tJvCA8zw.

3. Emily Reynolds, "The Agony of Sophia, the World's First Robot Citizen Condemned to a Lifeless Career in Marketing," *Wired*, January 6, 2018, https://www.wired.co.uk/article/sophia-robot-citizen-womens-rights-detriot-become-human-hanson-robotics.

4. Manfred Clynes and Nathan S. Kline, "Cyborgs and Space," *Astronautics* (September 1960): 27.

5. Norbert Wiener, *The Human Use of Human Beings: Cybernetics and Society*, 2nd ed. (Boston: Da Capo Press, 1954).

6. W. Ross Ashby, *An Introduction to Cybernetics* (London: Chapman and Hall Ltd., 1957).

7. Claude E. Shannon, "A Mathematical Theory of Communication," *Bell System Technical Journal* 27 (July 1948): 379–423, doi:10.1145/584091.584093.

8. Gregory Bateson, *Steps to an Ecology of Mind* (Chicago: University of Chicago Press, 1972).

9. Ronald Kline, "Where Are the Cyborgs in Cybernetics?," *Social Studies of Science* 39, no. 3 (2009): 331–362, doi:10.1177/0306312708101046.

10. Ashley Shew, "Disabled People in Space—Becoming Interplanetary," *Technology and Disability*, October 14, 2018, https://techanddisability.com/2018/10/14/disabled-people-in-space-becoming-interplanetary/.

11. Donna Haraway, "A Manifesto for Cyborgs: Science, Technology, and Socialist Feminism in the 1980s," *Socialist Review* 15, no. 2 (1985): 65–107.

12. Nina Lykke and Rosi Braidotti, eds., *Between Monsters, Goddesses and Cyborgs: Feminist Confrontations with Science, Medicine and Cyberspace* (London: Zed Books Ltd., 1996).

13. Chela Sandoval, "New Sciences: Cyborg Feminism and the Methodology of the Oppressed," in *The Cybercultures Reader*, ed. David Bell and Barbara M. Kennedy (New York: Routledge, 2000), 374–387.

14. Susan Leigh Star, "Power, Technology and the Phenomenology of Conventions: On Being Allergic to Onions," *Sociological Review* 38, no. S1 (2006): 26–56, doi:10.1111/j.1467–954X.1990.tb03347.x.

15. Meredith Broussard, *Artificial Unintelligence: How Computers Misunderstand the World* (Cambridge, MA: MIT Press, 2018).

Chapter 2

1. Ruth Schwartz Cowan, *More Work for Mother* (New York: Basic Books, 1983).

2. Yolande Strengers and Jenny Kennedy, *The Smart Wife: Why Siri, Alexa, and Other Smart Home Devices Need a Feminist Reboot* (Cambridge, MA: MIT Press, 2021).

3. Liz W. Faber, *The Computer's Voice: From Star Trek to Siri* (Minneapolis: University of Minnesota Press, 2020), 141.

4. Bruno Latour, "Where Are the Missing Masses? A Sociology of Few Mundane Objects," in *Shaping Technology/Building Society: Studies in Sociotechnical Change*, ed. Wiebe E. Bijker and John Law (Cambridge, MA: MIT Press, 1992), 151–180.

5. See https://www.oxfordlearnersdictionaries.com/us/definition/american _english/luddite, accessed July 15, 2019; http://www.nationalarchives.gov.uk /education/politics/g3/, accessed July 15, 2019.

6. See https://www.lexico.com/en/definition/taylorism, accessed July 15, 2019.

7. Social scientists today in a range of fields from anthropology, sociology, psychology, and organizational behavior have a long history of studying the contexts around humans and machines. One of the best-known studies, Julian Orr's *Talking about Machines*, used ethnographic methods in order to chronicle the practices of technical repair workers around Xerox machines. Similarly, Lucy Suchman's pioneering work at Xerox PARC around "human-machine configurations" offers a critical analysis of the ways in which the linear "plans" of an organization that are embedded in their design and use of technologies do not match up with the "situated actions" of people. Hamid Ekbia and Bonnie Nardi's book *Heteromation* traces the development of technology in tandem with the larger economic system of capitalism. Melissa Gregg's *Counterproductive* traces the ways in which productivity has been constructed as a goal that dominates both personal and professional time management. Julian Orr, *Talking about Machines: An Ethnography of a Modern Job* (Ithaca, NY: Cornell University Press, 2016); Lucy Suchman, *Plans and Situated Actions* (Cambridge: Cambridge University Press, 1987); Hamid R. Ekbia and Bonnie A. Nardi,

Heteromation, and Other Stories of Computing and Capitalis (Cambridge, MA: MIT Press, 2017); Melissa Gregg, *Counterproductive: Time Management in the Knowledge Economy* (Durham, NC: Duke University Press, 2018).

8. See Bill Vandenberg, "Is This the End of Jobs as We Know Them?," Open Society Foundations, April 7, 2015, https://www.opensocietyfoundations.org /voices/end-jobs-we-know-them.

9. Lilly Irani, "Justice for 'Data Janitors,'" *Public Books*, January 15, 2015, https://www.publicbooks.org/justice-for-data-janitors/.

10. Miriam Posner, "See No Evil," *Logic*, April 1, 2018.

11. Jean Yves Chainon and Kaitlyn Mullin, "Robots and Humans Team Up at Amazon," *New York Times*, September 10, 2017.

12. While as of 2017, the introduction of robots did not lead to job cuts, it is not clear that this will be the case in the foreseeable future.

13. Khari Johnson, "Amazon's 'Safe' New Robot Won't Fix Its Worker Injury Problem," *Wired*, July 8, 2022.

14. Laura Voss, *More Than Machines? The Attribution of (In)Animacy to Robot Technology* (Bielefeld, Germany: transcript, 2021).

15. Mary L. Gray and Siddharth Suri, *Ghost Work: How to Stop Silicon Valley from Building a New Global Underclass* (New York: Houghton Mifflin Harcourt, 2019); Alex Rosenblat, *Uberland: How Algorithms Are Rewriting the Rules of Work* (Berkeley: University of California Press, 2018); Brooke Erin Duffy, *(Not) Getting Paid to Do What You Love: Gender, Social Media, and Aspirational Work* (New Haven, CT: Yale University Press, 2017).

16. Noam Scheiber, "Inside an Amazon Warehouse, Robots' Ways Rub Off on Humans," *New York Times*, July 3, 2019.

17. Maggie Astor, "Microchip Implants for Employees? One Company Says Yes," *New York Times*, July 25, 2017; Haley Weiss, "Why You're Probably Getting a Microchip Implant Someday," *Atlantic*, September 21, 2018.

18. Steven Greenhouse, "'The Success Is Inspirational': The Fight for $15 Movement 10 Years On," *Guardian*, November 23, 2022.

19. Michael Sainato, "Whole Foods Workers Say Conditions Deteriorated after Amazon Takeover," *Guardian*, July 16, 2019.

20. Nitasha Tiku, "The Year Tech Workers Realized They Were Workers," *Wired*, December 24, 2018.

21. Lilly Irani and M. Six Silberman, "From Critical Design to Critical Infrastructure: Lessons from Turkopticon," *Interactions* 21, no. 4 (2014): 32–35.

22. Trebor Scholz, *Platform Cooperativism: Challenging the Corporate Sharing Economy* (New York: Rosa Luxemburg Foundation, 2016).

Chapter 3

1. Laura Forlano, "Data Rituals in Intimate Infrastructures: Crip Time and the Disabled Cyborg Body as an Epistemic Site of Feminist Science," *Catalyst: Feminism, Theory, Technoscience* 3, no. 2 (2017), doi:10.28968/cftt.v3i2.28843.

2. Carrie Rentschler and Benjamin Nothwehr, "Transmitting Insulin: The Design and Look of Insulin Delivery Devices as Technologies of Communication," *Catalyst: Feminism, Theory, Technoscience* 7, no. 1 (2021), doi:10.28968/cftt.v7i1.34567.

3. Michelle Murphy, "Immodest Witnessing: The Epistemology of Vaginal Self-Examination in the U.S. Feminist Self-Help Movement," *Feminist Studies* 30, no. 1 (2004): 115–147.

4. Laura Forlano, "Living Intimately with Machines: Can AI Be Disabled?," *Interactions* 30, no. 1 (January–February 2023): 3; Forlano, "Data Rituals in Intimate Infrastructures," 3.

5. Forlano, "Data Rituals in Intimate Infrastructures," 13.

6. Forlano, "The Danger of Intimate Algorithms," *Public Books*, April 13, 2020, https://www.publicbooks.org/the-danger-of-intimate-algorithms/.

7. Legacy Russell, *Glitch Feminism: A Manifesto* (New York: Verso Books, 2020), 12.

8. Nelly Oudshoorn, "Sustaining Cyborgs: Sensing and Tuning Agencies of Pacemakers and Implantable Cardioverter Defibrillators," *Social Studies of Science* 45, no. 1 (2015): 56–76, doi:10.1177/0306312714557377.

9. Erling Björgvinsson, Pelle Ehn, and Per-Anders Hillgren, "Design Things and Design Thinking: Contemporary Participatory Design Challenges," *Design Issues* 28, no. 3 (2012): 101–116.

10. Oudshoorn, "Sustaining Cyborgs," 69.

11. Mallory Kay Nelson, Ashley Shew, and Bethany Stevens, "Transmobility: Rethinking the Possibilities in Cyborg (Cripborg) Bodies," *Catalyst: Feminism, Theory, Technoscience* 5, no. 1 (2019): 3.

12. Nelson, Shew, and Stevens, "Transmobility," 4, 3.

13. As of this writing, Weise indicates "cy" as cy's preferred pronoun on public social media profiles.

14. Jillian Weise, "Dawn of the Tryborg," *New York Times*, November 30, 2016, https://www.nytimes.com/2016/11/30/opinion/the-dawn-of-the-tryborg.html.

15. Weise, "Dawn of the Tryborg."

16. Bureau of Labor Statistics, "Persons with a Disability: Labor Force Characteristics—2020," February 24, 2021, https://www.bls.gov/news.release/disabl.nr0.htm.

17. Bess Williamson, *Accessible America: A History of Disability and Design* (New York: NYU Press, 2019).

18. Richard W. Wertz and Dorothy C. Wertz, *Lying-In: A History of Childbirth in America*, exp. ed. (New Haven, CT: Yale University Press, 1989).

19. Wendy Kline, *Coming Home: How Midwives Changed Birth* (New York: Oxford University Press, 2019).

20. Murphy, "Immodest Witnessing."

21. Sophie Lewis, *Full Surrogacy Now: Feminism against Family* (New York: Verso Books, 2019).

22. Marie Jenkins Schwartz, *Birthing a Slave: Motherhood and Medicine in the Antebellum South* (Cambridge, MA: Harvard University Press, 2006).

23. Neda Atanasoski and Kalindi Vora, *Surrogate Humanity: Race, Robots, and the Politics of Technological Futures* (Durham, NC: Duke University Press, 2019).

24. Julia R. DeCook, "A [White] Cyborg's Manifesto: The Overwhelmingly Western Ideology Driving Technofeminist Theory," *Media, Culture, and Society* 43, no. 6 (2021): 1–10, doi:10.1177/0163443720957891.

25. Christina Sharpe, *In the Wake: On Blackness and Being* (Durham, NC: Duke University Press, 2016).

26. Alondra Nelson, "Future Texts," *Social Text* 20, no. 2 (2002): 1–16; Ruha Benjamin, "Racial Fictions, Biological Facts: Expanding the Sociological Imagination through Speculative Methods," *Catalyst: Feminism, Theory, Technoscience* 2, no. 2 (2016): 1–28, https://doi.org/10.28968/cftt.v2i2.28798; Sami Schalk, *Bodyminds Reimagined: (Dis)ability, Race, and Gender in Black Women's Speculative Fiction* (Durham, NC: Duke University Press, 2018).

Chapter 4

1. Legacy Russell, *Glitch Feminism: A Manifesto* (New York: Verso Books, 2020).

2. The opera was based on short stories by E. T. A. Hoffmann.

3. Thanks to scholar Sarah Dillon for hosting the Histories of Artificial Intelligence Reading Group on February 17, 2021.

4. Walter Gropius, "Bauhaus Manifesto 1919," Historic New England, https://gropius.house/location/bauhaus-manifesto/, accessed April 18, 2022.

5. Sook-Kyung Lee, Rudolf Frieling, John G. Hanhardt, Rachel Jans, Susanne Neuburger, and David Toop, eds., *Nam June Paik* (London: Tate Publishing, 2019), 14.

6. Lee et al., *Nam June Paik*, 9, 17.

7. Lee et al., *Nam June Paik*, 9, 9.

8. Nam June Paik, quoted in *Nam June Paik*, ed. Sook-Kyung Lee, Rudolf Frieling, John G. Hanhardt, Rachel Jans, Susanne Neuburger, and David Toop (London: Tate Publishing, 2019), 154.

9. Ron Labaco, *Out of Hand: Materializing the Post Digital*, Museum of Arts and Design, http://madmuseum.org/exhibition/out-hand#, accessed May 11, 2023.

10. Paul Dourish, *The Stuff of Bits: An Essay on the Materialities of Information* (Cambridge, MA: MIT Press, 2017).

11. Paul Vanouse and Tina Rivers Ryan, cur., *Difference Machines: Technology and Identity in Contemporary Art*, Buffalo AKG Art Museum, https://www.albrightknox.org/art/exhibitions/difference-machines-technology-and-identity-contemporary-art, accessed April 15, 2022.

12. Vanouse and Ryan, *Difference Machines*.

13. Kasperi Mäki-Reinikka, "Sensing Machines in Artistic Practice," in *Art as We Don't Know It*, ed. Erich Berger, Kasperi Mäki-Reinikka, Kira O'Reilly, and Helena Sederholm (Espoo, Finland: Essi Viitanen Aalto ARTS Books, 2020), 108–119.

14. Michelle Z. Donahue, "How a Color-Blind Artist Became the World's First Cyborg," *National Geographic*, April 3, 2017, https://www.nationalgeographic.com/science/article/worlds-first-cyborg-human-evolution-science.

15. Mäki-Reinikka, "Sensing Machines in Artistic Practice," 112.

16. Cyborg Foundation, https://www.cyborgfoundation.com, accessed January 1, 2023.

17. Mäki-Reinikka, "Sensing Machines in Artistic Practice," 115.

18. Stelarc, *RE-WIRED/RE-MIXED: Event for Dismembered Body*, http://stelarc.org/_activity-20353.php, accessed May 4, 2023.

19. Theresa Schubert, *mEat me*, https://www.theresaschubert.com/works/meat-me/, accessed December 31, 2022.

20. Kaethe Wenzel, *Bone Bots*, http://www.kaethewenzel.de/html/englbonebot2.htm, accessed April 15, 2022.

21. Andrew Bolton, *Rei Kawakubo / Comme des Garçons: Art of the In-between* (New York: Metropolitan Museum of Art, 2017), 12.

22. Bolton, *Rei Kawakubo / Comme des Garçons*, 13.

23. Bolton, *Rei Kawakubo / Comme des Garçons*, 14–15.

24. Uwe Schütte, *Kraftwerk: Future Music from Germany* (New York: Penguin Books, 2020), 114, 272.

25. Lindsay Zoladz, "Amplifying the Women Who Pushed Synthesizers into the Future," *New York Times*, April 21, 2021; Steve Marshall, "The Story of the

BBC Radiophonic Workshop," *SOS*, April 2008, https://www.soundonsound
.com/people/story-bbc-radiophonic-workshop.

26. Thanks to Sinnreich, author of *Mashed Up*, a book about remix and "configurable culture," for the conversation on February 17, 2021, that helped trace this history.

27. Kodwo Eshun, *More Brilliant than the Sun: Adventures in Sonic Fiction* (New York: Verso Books, 2018), 82, 125.

28. Marcus Barnes, "Mysteries of the Deep: How Drexciya Reimagined Slavery to Create an Afrofuturist Utopia," *Mixmag*, October 19, 2020.

29. Helen Scales, "Drexciya: How Afrofuturism Is Inspiring Calls for an Ocean Memorial to Slavery," *Guardian*, January 25, 2021.

30. Katherine McKittrick, *Dear Science and Other Stories* (Durham, NC: Duke University Press, 2021), 57.

31. Martine Syms, "The Mundane Afrofuturist Manifesto," in *Colored People Time*, ed. Meg Onli and Amber Rose Johnson (Philadelphia: University of Pennsylvania, 2019), 101–116.

32. VNS Matrix, "The Cyberfeminist Manifesto for the 21st Century," 1991, https://vnsmatrix.net/projects/the-cyberfeminist-manifesto-for-the-21st-century.

33. VNS Matrix, "The Cyberfeminist Manifesto."

34. VNS Matrix, "The Cyberfeminist Manifesto."

35. Mindy Seu, *Cyberfeminism Index*, https://cyberfeminismindex.com/about/, accessed January 28, 2021.

36. Marie Hoejlund, "Sharing as Survival: Mindy Seu on the Cyberfeminism Index," *Gradient*, November 9, 2020.

37. Lisa Phillips, "Foreword," in *Lee Hershman Leeson: Twisted*, ed. Margot Norton (New York: New Museum, 2021), 7.

38. Karen Archey, "Dead and Alive," in *Lee Hershman Leeson: Twisted*, ed. Margot Norton (New York: New Museum, 2021), 30–41.

39. Quoted in Margot Norton, ed., *Lee Hershman Leeson: Twisted* (New York: New Museum, 2021).

40. Lee Bul, "A Feeling about Freedom: Stephanie Rosenthal in Conversation with Lee Bul," in *Lee Bul*, ed. Stephanie Rosenthal (Cologne: Walther Konig, 2018), 87.

41. Bul, "A Feeling about Freedom," 87.

42. Bul, "A Feeling about Freedom," 88.

43. Stephanie Rosenthal, "A Feeling about Freedom: Stephanie Rosenthal in conversation with Lee Bul," in *Lee Bul*, ed. Stephanie Rosenthal (Cologne: Walther Konig, 2018), 88.

44. Thanks to the helpful conversation with artist Laura Splan on January 27, 2021, for this description.

45. Stephanie Rosenthal, "Lee Bul: Crashing," in *Lee Bul*, ed. Stephanie Rosenthal (Cologne: Walther Konig, 2018), 12.

Chapter 5

1. Donna Haraway, "A Manifesto for Cyborgs: Science, Technology, and Socialist Feminism in the 1980s," *Socialist Review* 15, no. 2 (1985): 66.

2. Manfred Clynes and Nathan S. Kline, "Cyborgs and Space," *Astronautics* (September 1960): 26–27, 74–75; Ronald Kline, "Where Are the Cyborgs in Cybernetics?," *Social Studies of Science* 39, no. 3 (2009): 331–362, doi:10.1177/0306312708101046.

3. Norbert Wiener, *The Human Use of Human Beings: Cybernetics and Society*, 2nd ed. (Boston: Da Capo Press, 1954).

4. Clynes and Kline, "Cyborgs and Space," 27.

5. Haraway, "A Manifesto for Cyborgs," 65.

6. Haraway, "A Manifesto for Cyborgs," 66.

7. Haraway, "A Manifesto for Cyborgs," 70.

8. Mar Hicks, *Programmed Inequality: How Britain Discarded Women Technologists and Lost Its Edge in Computing* (Cambridge, MA: MIT Press, 2017); Jennifer S. Light, "When Computers Were Women," *Technology and Culture* 40, no. 3 (1999): 455–483.

9. Haraway, "A Manifesto for Cyborgs," 90, 85.

10. Haraway, "A Manifesto for Cyborgs," 85–86.

11. Haraway, "A Manifesto for Cyborgs," 86.

12. Haraway, "A Manifesto for Cyborgs," 66, 67.

13. Haraway, "A Manifesto for Cyborgs," 68, 69, 70.

14. Jillian Weise, "Common Cyborg," *Granta*, September 24, 2018, https://granta.com/common-cyborg/.

15. Haraway, "A Manifesto for Cyborgs," 75.

16. Haraway, "A Manifesto for Cyborgs," 72.

17. Julia R. DeCook, "A [White] Cyborg's Manifesto: The Overwhelmingly Western Ideology Driving Technofeminist Theory," *Media, Culture, and Society* 43, no. 6 (2021): 1–10, doi:10.1177/0163443720957891.

18. Sherry B. Ortner, "Is Female to Male as Nature Is to Culture?," *Feminist Studies* 1, no. 2 (Autumn 1972): 5–31; Sandra Harding, *The Science Question in Feminism* (Ithaca, NY: Cornell University Press, 1986); Evelyn Fox Keller, *A Feeling for the Organism: The Life and Work of Barbara McClintock* (New York:

Owl Books, 2003); Emily Martin, *The Woman in the Body: A Cultural Analysis of Reproduction* (Boston: Beacon Press, 1987).

19. Haraway, "A Manifesto for Cyborgs," 72.

Chapter 6

1. Alison Kafer, *Feminist, Queer, Crip* (Bloomington: Indiana University Press, 2013).

2. Donna J. Haraway, *Staying with the Trouble: Making Kin in the Chthulucene* (Durham, NC: Duke University Press, 2016), 1.

3. See "Elon Musk's Neuralink Chip Tested Live in Pig Brains," CNET, August 28, 2020, https://www.youtube.com/watch?v=NqbQuZOFvOQ.

4. Eliza Strickland and Mark Harris, "Their Bionic Eyes Are Now Obsolete and Unsupported," *IEEE Spectrum*, February 15, 2022.

5. Julia R. DeCook, "A [White] Cyborg's Manifesto: The Overwhelmingly Western Ideology Driving Technofeminist Theory," *Media, Culture, and Society* 43, no. 6 (2021): 1–3, doi:10.1177/0163443720957891.

6. DeCook, "A [White] Cyborg's Manifesto," 6.

7. Alexander G. Weheliye, *Habeas Viscus: Racializing Assemblages, Biopolitics, and Black Feminist Theories of the Human* (Durham, NC: Duke University Press, 2014), 8.

8. Neda Atanasoski and Kalindi Vora, *Surrogate Humanity: Race, Robots, and the Politics of Technological Futures* (Durham, NC: Duke University Press, 2019), 5.

9. bell hooks, "Theory as Liberatory Practice," *Yale Journal of Law and Feminism* 4 (1991): 1–2.

10. Paulo Freire, *Pedagogy of the Oppressed* (New York: Bloomsbury, 2018), 51.

11. Beth Coleman, "Race as Technology," *Camera Obscura: Feminism, Culture, and Media Studies* 24, no. 1 (70) (2009): 177–207; Wendy Hui Kyong Chun, "Race and/as Technology, or How to Do Things to Race," in *Race after the Internet*, ed. Lisa Nakamura and Peter A. Chow-White (New York: Routledge, 2013), 44–66.

12. Jasmine Johnson, "The #OptimisticChallenge," in *Black Futures*, ed. Kimberly Drew and Jenna Wortham (New York: One World, 2021), 154.

13. Kafer, *Feminist, Queer, Crip*, 105.

14. Kafer, *Feminist, Queer, Crip*, 106.

15. Ashley Shew, "Ableism, Technoableism, and Future AI," *IEEE Technology and Society Magazine* 39, no. 1 (2020): 43.

16. Jennifer Michalowski, "New Bionics Center Established at MIT with $24 Million Gift," *MIT News*, September 23, 2021, https://news.mit.edu/2021/new -bionics-center-established-mit-24-million-gift-0923.

17. According to a conversation on Twitter on September 24, 2021. See Dr. Laura Forlano (@laura4lano), "This is (potentially) exciting but the framing is very ableist," Twitter, September 24, 2021, 8:29 a.m., https://twitter.com /laura4lano/status/1441379361068773394.

18. Kafer, *Feminist, Queer, Crip*, 116.

19. Liz Jackson, Alex Haagaard, and Rua Williams, "Disability Dongle," *Platypus: The CASTAC Blog*, April 19, 2022, https://blog.castac.org/2022/04 /disability-dongle/.

20. Personal correspondence with Louis Hickman on February 7, 2022.

21. Kafer, *Feminist, Queer, Crip*, 128–129.

BIBLIOGRAPHY

Archey, Karen. "Dead and Alive." In *Lee Hershman Leeson: Twisted*, edited by Margot Norton, 30–41. New York: New Museum, 2021.

Ashby, W. Ross. *An Introduction to Cybernetics*. London: Chapman and Hall Ltd., 1957.

Astor, Maggie. "Microchip Implants for Employees? One Company Says Yes." *New York Times*, July 25, 2017.

Atanasoski, Neda, and Kalindi Vora. *Surrogate Humanity: Race, Robots, and the Politics of Technological Futures*. Durham, NC: Duke University Press, 2019.

Barnes, Marcus. "Mysteries of the Deep: How Drexciya Reimagined Slavery to Create an Afrofuturist Utopia." *Mixmag*, October 19, 2020.

Bateson, Gregory. *Steps to an Ecology of Mind*. Chicago: University of Chicago Press, 1972.

Benjamin, Ruha. "Racial Fictions, Biological Facts: Expanding the Sociological Imagination through Speculative Methods." *Catalyst: Feminism, Theory, Technoscience* 2, no. 2 (2016): 1–28. https://doi.org/10.28968/cftt.v2i2.28798.

Björgvinsson, Erling, Pelle Ehn, and Per-Anders Hillgren. "Design Things and Design Thinking: Contemporary Participatory Design Challenges." *Design Issues* 28, no. 3 (2012): 101–116.

Bolton, Andrew. *Rei Kawakubo / Comme des Garçons: Art of the In-between*. New York: Metropolitan Museum of Art, 2017.

Broussard, Meredith. *Artificial Unintelligence: How Computers Misunderstand the World*. Cambridge, MA: MIT Press, 2018.

Bureau of Labor Statistics. "Persons with a Disability: Labor Force Characteristics—2020." February 24, 2021. https://www.bls.gov/news.release/disabl.nr0.htm.

Chainon, Jean Yves, and Kaitlyn Mullin. "Robots and Humans Team Up at Amazon." *New York Times*, September 10, 2017.

Chun, Wendy Hui Kyong. "Race and/as Technology, or How to Do Things to Race." In *Race after the Internet*, edited by Lisa Nakamura and Peter A. Chow-White, 44–66. New York: Routledge, 2013.

Clynes, Manfred, and Nathan S. Kline. "Cyborgs and Space." *Astronautics* (September 1960): 26–27, 74–75.

Coleman, Beth. "Race as Technology." *Camera Obscura: Feminism, Culture, and Media Studies* 24, no. 1 (70) (2009): 177–207.

Cowan, Ruth Schwartz. *More Work for Mother*. New York: Basic Books, 1983.

DeCook, Julia R. "A [White] Cyborg's Manifesto: The Overwhelmingly Western Ideology Driving Technofeminist Theory." *Media, Culture, and Society* 43, no. 6 (2020): 1–10. doi:10.1177/0163443720957891.

Donahue, Michelle Z. "How a Color-Blind Artist Became the World's First Cyborg." *National Geographic*, April 3, 2017. https://www.nationalgeographic .com/science/article/worlds-first-cyborg-human-evolution-science.

Dourish, Paul. *The Stuff of Bits: An Essay on the Materialities of Information*. Cambridge, MA: MIT Press, 2017.

Duffy, Brooke Erin. *(Not) Getting Paid to Do What You Love: Gender, Social Media, and Aspirational Work*. New Haven, CT: Yale University Press, 2017.

Ekbia, Hamid R., and Bonnie A. Nardi. *Heteromation, and Other Stories of Computing and Capitalism*. Cambridge, MA: MIT Press, 2017.

Eshun, Kodwo. *More Brilliant than the Sun: Adventures in Sonic Fiction*. New York: Verso Books, 2018.

Faber, Liz W. *The Computer's Voice: From Star Trek to Siri*. Minneapolis: University of Minnesota Press, 2020.

Forlano, Laura. "The Danger of Intimate Algorithms." *Public Books*, April 13, 2020. https://www.publicbooks.org/the-danger-of-intimate-algorithms/.

Forlano, Laura. "Data Rituals in Intimate Infrastructures: Crip Time and the Disabled Cyborg Body as an Epistemic Site of Feminist Science." *Catalyst: Feminism, Theory, Technoscience* 3, no. 2 (2017). doi:10.28968/cftt.v3i2.28843.

Forlano, Laura. "Living Intimately with Machines: Can AI Be Disabled?" *Interactions* 30, no. 1 (January–February 2023): 24–29.

Freire, Paulo. *Pedagogy of the Oppressed*. New York: Bloomsbury, 2018.

Gray, Mary L., and Siddharth Suri. *Ghost Work: How to Stop Silicon Valley from Building a New Global Underclass*. New York: Houghton Mifflin Harcourt, 2019.

Greenhouse, Steven. "'The Success Is Inspirational': The Fight for $15 Movement 10 Years On." *Guardian*, November 23, 2022.

Gregg, Melissa. *Counterproductive: Time Management in the Knowledge Economy*. Durham, NC: Duke University Press, 2018.

Haraway, Donna. "A Manifesto for Cyborgs: Science, Technology, and Socialist Feminism in the 1980s." *Socialist Review* 15, no. 2 (1985): 65–107.

Haraway, Donna J. *Staying with the Trouble: Making Kin in the Chthulucene*. Durham, NC: Duke University Press, 2016.

Harding, Sandra. *The Science Question in Feminism*. Ithaca, NY: Cornell University Press, 1986.

Hicks, Mar. *Programmed Inequality: How Britain Discarded Women Technologists and Lost Its Edge in Computing*. Cambridge, MA: MIT Press, 2017.

Hoejlund, Marie. "Sharing as Survival: Mindy Seu on the Cyberfeminism Index." *The Gradient*, November 9, 2020.

hooks, bell. "Theory as Liberatory Practice." *Yale Journal of Law and Feminism* 4 (1991): 1–2.

Irani, Lilly. "Justice for 'Data Janitors.'" *Public Books*, January 15, 2015. https://www.publicbooks.org/justice-for-data-janitors/.

Irani, Lilly, and M. Six Silberman. "From Critical Design to Critical Infrastructure: Lessons from Turkopticon." *Interactions* 21, no. 4 (2014): 32–35.

Jackson, Liz, Alex Haagaard, and Rua Williams. "Disability Dongle." *Platypus: The CASTAC Blog*, April 19, 2022.

Johnson, Jasmine. "The #OptimisticChallenge." In *Black Futures*, edited by Kimberly Drew and Jenna Wortham, 152–155. New York: One World, 2021.

Johnson, Khari. "Amazon's 'Safe' New Robot Won't Fix Its Worker Injury Problem." *Wired*, July 8, 2022.

Kafer, Alison. *Feminist, Queer, Crip*. Bloomington: Indiana University Press, 2013.

Keller, Evelyn Fox. *A Feeling for the Organism: The Life and Work of Barbara McClintock*. New York: Owl Books, 2003.

Kline, Ronald. "Where Are the Cyborgs in Cybernetics?" *Social Studies of Science* 39, no. 3 (2009): 331–362. doi:10.1177/0306312708101046.

Kline, Wendy. *Coming Home: How Midwives Changed Birth*. New York: Oxford University Press, 2019.

Labaco, Ron. *Out of Hand: Materializing the Post Digital*. Museum of Arts and Design. Accessed May 11, 2023. http://madmuseum.org/exhibition/out-hand#.

Latour, Bruno. "Where Are the Missing Masses? A Sociology of Few Mundane Objects." In *Shaping Technology/Building Society: Studies in Sociotechnical Change*, edited by Wiebe E. Bijker and John Law, 151–180. Cambridge, MA: MIT Press, 1992.

Lee, Sook-Kyung, Rudolf Frieling, John G. Hanhardt, Rachel Jans, Susanne Neuburger, and David Toop, eds. *Nam June Paik*. London: Tate Publishing, 2019.

Lewis, Sophie. *Full Surrogacy Now: Feminism against Family*. New York: Verso Books, 2019.

Light, Jennifer S. "When Computers Were Women." *Technology and Culture* 40, no. 3 (1999): 455–483.

Lykke, Nina, and Rosi Braidotti, eds. *Between Monsters, Goddesses and Cyborgs: Feminist Confrontations with Science, Medicine and Cyberspace*. London: Zed Books Ltd., 1996.

Mäki-Reinikka, Kasperi. "Sensing Machines in Artistic Practice." In *Art as We Don't Know It*, edited by Erich Berger, Kasperi Mäki-Reinikka, Kira O'Reilly, and Helena Sederholm, 108–119. Espoo, Finland: Essi Viitanen Aalto ARTS Books, 2020.

McKittrick, Katherine. *Dear Science and Other Stories*. Durham, NC: Duke University Press, 2021.

Murphy, Michelle. "Immodest Witnessing: The Epistemology of Vaginal Self-Examination in the U.S. Feminist Self-Help Movement." *Feminist Studies* 30, no. 1 (2004): 115–147.

Nelson, Alondra. "Future Texts." *Social Text* 20, no. 2 (2002): 1–16.

Nelson, Mallory Kay, Ashley Shew, and Bethany Stevens. "Transmobility: Rethinking the Possibilities in Cyborg (Cripborg) Bodies." *Catalyst: Feminism, Theory, Technoscience* 5, no. 1 (2019): 1–20. doi:10.28968/cftt.v5i1.29617.

Norton, Margot, ed. *Lee Hershman Leeson: Twisted*. New York: New Museum, 2021.

Orr, Julian E. *Talking about Machines: An Ethnography of a Modern Job*. Ithaca, NY: Cornell University Press, 2016.

Ortner, Sherry B. "Is Female to Male as Nature Is to Culture?" *Feminist Studies* 1, no. 2 (Autumn 1972): 5–31.

Oudshoorn, Nelly. "Sustaining Cyborgs: Sensing and Tuning Agencies of Pacemakers and Implantable Cardioverter Defibrillators." *Social Studies of Science* 45, no. 1 (2015): 56–76. doi:10.1177/0306312714557377.

Phillips, Lisa. "Foreword." In *Lee Hershman Leeson: Twisted*, edited by Margot Norton, 6–11. New York: New Museum, 2021.

Posner, Miriam. "See No Evil." *Logic*, April 1, 2018.

Rentschler, Carrie, and Benjamin Nothwehr. "Transmitting Insulin: The Design and Look of Insulin Delivery Devices as Technologies of Communication." *Catalyst: Feminism, Theory, Technoscience* 7, no. 1 (2021). doi:10.28968/cftt .v7i1.34567.

Reynolds, Emily. "The Agony of Sophia, the World's First Robot Citizen Condemned to a Lifeless Career in Marketing." *Wired*, January 1, 2018. https://www.wired.co.uk/article/sophia-robot-citizen-womens-rights-detriot-become -human-hanson-robotics.

Rosenblat, Alex. *Uberland: How Algorithms Are Rewriting the Rules of Work*. Berkeley: University of California Press, 2018.

Rosenthal, Stephanie, ed. *Lee Bul*. Cologne: Walther Konig, 2018.

Russell, Legacy. *Glitch Feminism: A Manifesto*. New York: Verso Books, 2020.

Sainato, Michael. "Whole Foods Workers Say Conditions Deteriorated after Amazon Takeover." *Guardian*, July 16, 2019.

Sandoval, Chela. "New Sciences: Cyborg Feminism and the Methodology of the Oppressed." In *The Cybercultures Reader*, edited by David Bell and Barbara M. Kennedy, 374–387. New York: Routledge, 2000.

Scales, Helen. "Drexciya: How Afrofuturism Is Inspiring Calls for an Ocean Memorial to Slavery." *Guardian*, January 25, 2021.

Schalk, Sami. *Bodyminds Reimagined: (Dis)ability, Race, and Gender in Black Women's Speculative Fiction*. Durham, NC: Duke University Press, 2018.

Scheiber, Noam. "Inside an Amazon Warehouse, Robots' Ways Rub Off on Humans." *New York Times*, July 3, 2019.

Scholz, Trebor. *Platform Cooperativism: Challenging the Corporate Sharing Economy*. New York: Rosa Luxemburg Foundation, 2016.

Schütte, Uwe. *Kraftwerk: Future Music from Germany*. New York: Penguin Books, 2020.

Schwartz, Marie Jenkins. *Birthing a Slave: Motherhood and Medicine in the Antebellum South*. Cambridge, MA: Harvard University Press, 2006.

Shannon, Claude E. "A Mathematical Theory of Communication." *Bell System Technical Journal* 27 (July 1948): 379–423. doi:10.1145/584091.584093.

Sharpe, Christina. *In the Wake: On Blackness and Being*. Durham, NC: Duke University Press, 2016.

Shew, Ashley. "Ableism, Technoableism, and Future AI." *IEEE Technology and Society Magazine* 39, no. 1 (2020): 40–85.

Shew, Ashley. "Disabled People in Space—Becoming Interplanetary." *Technology and Disability*, October 14, 2018. https://techanddisability.com/2018/10/14/disabled-people-in-space-becoming-interplanetary/.

Star, Susan Leigh. "Power, Technology and the Phenomenology of Conventions: On Being Allergic to Onions." *Sociological Review* 38, no. S1 (2006): 26–56. doi:10.1111/j.1467-954X.1990.tb03347.x.

Strengers, Yolande, and Jenny Kennedy. *The Smart Wife: Why Siri, Alexa, and Other Smart Home Devices Need a Feminist Reboot*. Cambridge, MA: MIT Press, 2021.

Strickland, Eliza, and Mark Harris. "Their Bionic Eyes Are Now Obsolete and Unsupported." *IEEE Spectrum*, February 15, 2022.

Suchman, Lucy. *Plans and Situated Actions*. Cambridge: Cambridge University Press, 1987.

Syms, Martine. "The Mundane Afrofuturist Manifesto." In *Colored People Time*, edited by Meg Onli and Amber Rose Johnson, 101–116. Philadelphia: University of Pennsylvania, 2019.

Tiku, Nitasha. "The Year Tech Workers Realized They Were Workers." *Wired*, December 24, 2018.

Voss, Laura. *More Than Machines? The Attribution of (In)Animacy to Robot Technology*. Bielefeld, Germany: transcript, 2021.

Weheliye, Alexander G. *Habeas Viscus: Racializing Assemblages, Biopolitics, and Black Feminist Theories of the Human*. Durham, NC: Duke University Press, 2014.

Weise, Jillian. "Common Cyborg." *Granta*, September 24, 2018. https://granta .com/common-cyborg/.

Weise, Jillian. "Dawn of the Tryborg." *New York Times*, November 30, 2016. https://www.nytimes.com/2016/11/30/opinion/the-dawn-of-the-tryborg .html.

Weiss, Haley. "Why You're Probably Getting a Microchip Implant Someday." *Atlantic*, September 21, 2018.

Wertz, Richard W., and Dorothy C. Wertz. *Lying-In: A History of Childbirth in America*. Exp. ed. New Haven, CT: Yale University Press, 1989.

Wiener, Norbert. *The Human Use of Human Beings: Cybernetics and Society*. 2nd ed. Boston: Da Capo Press, 1954.

Williamson, Bess. *Accessible America: A History of Disability and Design*. New York: NYU Press, 2019.

Zoladz, Lindsay. "Amplifying the Women Who Pushed Synthesizers into the Future." *New York Times*, April 21, 2021.

FURTHER READING

This section presents some additional resources to help you continue your journey of learning about cyborgs. We have compiled some of our favorite books, essays, movies, podcast episodes, and talks that can help deepen your engagement with the main text. For this list, we have prioritized items that are openly available or cheap and easy to find in a published form. We have included selections to accompany the five main chapters of the book, excluding the introduction and conclusion. Some of the items below appear in the main text, while others are things we find ourselves returning to enjoy over and over again that might offer extra inspiration after you finish this book.

For the full list of citations, which includes items that are behind journal paywalls and only available in more pricey scholarly books, please refer to the bibliography.

Chapter 2

Gray, Mary L., and Siddharth Suri. *Ghost Work: How to Stop Silicon Valley from Building a New Global Underclass*. New York: Houghton Mifflin Harcourt, 2019.

Gregg, Melissa. *Counterproductive: Time Management in the Knowledge Economy*. Durham, NC: Duke University Press, 2018.

Irani, Lilly. "Justice for 'Data Janitors.'" *Public Books*, January 15, 2015. https://www.publicbooks.org/justice-for-data-janitors/.

Rivera, Alex, dir. *Sleep Dealer*. Los Angeles: Maya Entertainment, 2008.

Rosenblat, Alex. *Uberland: How Algorithms Are Rewriting the Rules of Work*. Berkeley: University of California Press, 2018.

Salehi, Niloufar, Lilly Irani, Ali Al Khatib, and Michael Bernstein. "Dynamo: Designing Interactive Technology to Support Social Movements in Digital Labor." Future of Work Project, Open Society Foundations, 2014.

Scholz, Trebor. *Platform Cooperativism: Challenging the Corporate Sharing Economy*. New York: Rosa Luxemburg Foundation, 2016.

Strengers, Yolande, and Jenny Kennedy. *The Smart Wife: Why Siri, Alexa, and Other Smart Home Devices Need a Feminist Reboot*. Cambridge, MA: MIT Press, 2021.

Tarnoff, Ben. "From Manchester to Barcelona." *Logic Magazine*, December 7, 2019. https://logicmag.io/nature/from-manchester-to-barcelona/.

Wang, Xiaowei. *Blockchain Chicken Farm: And Other Stories of Tech in China's Countryside*. New York: FSG Originals X Logic, 2020.

Chapter 3

Bodies: The Complete Season. Podcast, Flash Forward, June 11, 2019. https://www.flashforwardpod.com/2019/06/11/bodies-the-complete-season/.

Cyborgs. Podcast, Disability Visibility, December 18, 2019. https://disabilityvisibilityproject.com/2019/12/18/ep-66-cyborgs/.

Forlano, Laura. "The Danger of Intimate Algorithms." *Public Books*, April 13, 2020. https://www.publicbooks.org/the-danger-of-intimate-algorithms/.

Hendren, Sara. *What Can a Body Do? How We Meet the Built World*. New York: Riverhead Books, 2020.

Lewis, Sophie. *Full Surrogacy Now: Feminism against Family*. New York: Verso Books, 2019.

Nelson, Mallory Kay, Ashley Shew, and Bethany Stevens. "Transmobility: Rethinking the Possibilities in Cyborg (Cripborg) Bodies." *Catalyst: Feminism, Theory, Technoscience* 5, no. 1 (2019): 1–20. doi:10.28968/cftt.v5i1.29617.

Schalk, Sami. *Bodyminds Reimagined: (Dis)ability, Race, and Gender in Black Women's Speculative Fiction*. Durham, NC: Duke University Press, 2018.

Sins Invalid. *Skin, Tooth and Bone: The Basis of Movement Is Our People, a Disability Justice Primer*. Berkeley, CA: Sins Invalid, 2019.

Weise, Jillian. "Dawn of the Tryborg." *New York Times*, November 30, 2016. https://www.nytimes.com/2016/11/30/opinion/the-dawn-of-the-tryborg.html.

Chapter 4

Art's Work in the Age of Biotechnology. Exhibition, University of Pittsburgh. Accessed March 24, 2023. https://www.artsworkintheageofbiotechnology.org.

Biodesigned. Accessed March 24, 2023. https://www.biodesigned.org.

Coogler, Ryan, dir. *Black Panther*. Burbank, CA: Marvel Studios, 2018.

Difference Machines. Exhibition, Buffalo AKG Art Museum. Accessed March 24, 2023. https://buffaloakg.org/difference-machines-resources.

Eshun, Kodwo. *More Brilliant than the Sun: Adventures in Sonic Fiction*. New York: Verso Books, 2018.

Mitchell, Robert E. *Bioart and the Vitality of Media*. Seattle: University of Washington Press, 2015.

Myers, William. *Bio Art: Altered Realities*. New York: Thames and Hudson, 2015.

Out of Hand: Materializing the Postdigital. Exhibition, Museum of Arts and Design, October 16, 2013–June 1, 2014. https://madmuseum.org/exhibition /out-hand.

Rei Kawakubo / Comme des Garçons: Art of the In-between. Exhibition, Metropolitan Museum of Art, 2017. https://www.metmuseum.org/exhibitions /listings/2017/rei-kawakubo/exhibition-galleries.

Seu, Mindy. *Cyberfeminism Index*. Los Angeles: Inventory Press, 2022. https:// cyberfeminismindex.com.

SK-Interfaces—Exploding Borders in Art, Technology and Society. Exhibition video, Vimeo, July 3, 2012. https://vimeo.com/45131984.

Talk to Me. Exhibition, Museum of Modern Art, New York, July 24–November 7, 2011. https://www.moma.org/interactives/exhibitions/2011/talktome/.

VNS Matrix / Merchants of Slime. Accessed March 24, 2023. https://vnsmatrix .net/.

Zylinska, Joanna. *The Cyborg Experiments: The Extensions of the Body in the Media Age*. New York: Bloomsbury, 2002.

Chapter 5

Glabau, Danya. "Do Cyborgs Have Politics?" *Pax Solaria*, 2017. http://www .paxsolaria.net/monthly/do-cyborgs-have-politics.

Haraway, Donna J. *Manifestly Haraway*. Minneapolis: University of Minnesota Press, 2016.

Harbisson, Neil. "I Listen to Color." TED, July 20, 2012. https://youtu.be/ ygRNoieAnzI.

Laboria Cuboniks. "Xenofeminism: A Politics for Alienation." Accessed March 24, 2023. https://laboriacuboniks.net/manifesto/xenofeminism-a-politics-for -alienation/.

Madrigal, Alexis C. "The Man Who First Said 'Cyborg,' 50 Years Later." *Atlantic*, September 30, 2010. https://www.theatlantic.com/technology/archive/2010/09/the-man-who-first-said-cyborg-50-years-later/63821/.

Russell, Legacy. *Glitch Feminism: A Manifesto*. New York: Verso Books, 2020.

Weise, Jillian. "Common Cyborg." *Granta*, September 24, 2018. https://granta.com/common-cyborg/.

Wiener, Norbert. *The Human Use of Human Beings: Cybernetics and Society*. 2nd ed. Boston: Da Capo Press, 1954.

Chapter 6

Atanasoski, Neda, and Kalindi Vora. *Surrogate Humanity: Race, Robots, and the Politics of Technological Futures*. Durham, NC: Duke University Press, 2019.

Benjamin, Ruha. *Race after Technology: Abolitionist Tools for the New Jim Code*. Cambridge, UK: Polity Press, 2019.

DeCook, Julia R. "A [White] Cyborg's Manifesto: The Overwhelmingly Western Ideology Driving Technofeminist Theory." *Media, Culture, and Society* 43, no. 6 (2020): 1–10. doi:10.1177/0163443720957891.

Jackson, Liz, Alex Haagaard, and Rua Williams. "Disability Dongle." *Platypus: The CASTAC Blog*, April 19, 2022. https://blog.castac.org/2022/04/disability-dongle/.

Jackson, Zakiyyah Iman. *Becoming Human: Matter and Meaning in an Antiblack World*. New York: NYU Press, 2020.

Kafer, Alison. *Feminist, Queer, Crip*. Bloomington: Indiana University Press, 2013.

Lewis, Jason Edward, Noelani Arista, Archer Pechawis, and Suzanne Kite. "Making Kin with the Machines." *Journal of Design and Science* (July 16, 2018). https://jods.mitpress.mit.edu/pub/lewis-arista-pechawis-kite/release/1.

Nolan, Jonathan, and Lisa Joy, creators. *Westworld*, season 1. New York: HBO, 2016.

Weheliye, Alexander G. *Habeas Viscus: Racializing Assemblages, Biopolitics, and Black Feminist Theories of the Human*. Durham, NC: Duke University Press, 2014.

Abe, Shuya, 88

Access washing, 159

Afrofuturism, 16–17, 101–104, 105–106, 153–154, 179

Agency, 38–39, 44, 64–70, 75–76, 101, 161

A.I. (film), 86

Algorithmic bosses, 38, 45–47

Amazon, 27, 28–35, 43, 46–47, 50–51

Ames, Morgan, 149

Androids (humanoid robots), 2, 169. *See also* Sophia (android)

Animacy, 44

Archey, Karen, 107

Artificial intelligence (AI), 4, 179

Artificial senses, 93

Art of the In-between (exhibition), 96–98

Ashby, W. Ross, 11

Astronautics (journal), 9–10, 115–116

Atanasoski, Neda, 151–152

Atlantic (magazine), 48

Aunt (Paik), 88

Automated insulin pumps, 62–63, 110–111

Automated teller machines (ATMs), 39–40

Automation (cyborg labor), 179
    algorithmic bosses and, 38, 45–47
    Amazon's role in, 28–35, 43, 46–47

biotech enhancement of workers and, 47–49

common narratives about, 17–18, 27–28, 52–53

cyborg culture and, 100–101

future of work and, 49–52, 173–174

gender and sexuality norms and, 35–37

remote and distributed work and, 41–45

replacement of human labor and, 28, 37–41

as represented in cinema, 31, 41, 45, 48–49, 86

Bambaataa, Afrika, 101

Barratt, Virginia, 104–105

Barrio, Itziar, 111

Bateson, Gregory, 11

Battlestar Galactica (television series), 86

Bauhaus, 86–88

Benjamin, Ruha, 17, 79, 149

Between Monsters, Goddesses, and Cyborgs (Lykke and Braidotti), 14–15

Bezos, Jeff, 17

Bioart, 92–96

Black cyberfeminism, 105–106

Black feminism, 77, 170

Black Futures (Johnson), 154

Black Mirror (television series), 142

Black Panther (film), 103, 154

Bladerunner (film), 86
Bone Bots (Wenzel), 95
Brain-machine interfaces, 140–142
Break dancing, 100–101
Breathing Machines (Leeson), 107
Broussard, Meredith, 18–19
Bul, Lee, 108–110
Bureau of Labor Statistics, 69
Butler, Octavia, 103

Cage, John, 88
Capitalism, 51–52
Care, 174–175
Care work, 42, 122–123
Cinema
    Afrofuturism and, 103, 154
    automation in, 31, 41, 45, 48–49,
        86
Clothes / Not Clothes (Kawakubo),
    99
Clynes, Manfred, 9–10, 115–116
Collins, Patricia Hill, 170
Colonialism, 127–128, 131, 138,
    146. *See also* Race and racism
"Common Cyborg" (Weise), 127
Computer's Voice, The (Faber), 36
Coogler, Ryan, 103
Counterproductive (Gregg), 184n7
COVID-19 pandemic, 29, 142–143,
    148, 157–158
Cowan, Ruth Schwartz, 36
Crenshaw, Kimberlé, 170
Cripborgs, 60, 66–68, 69–71, 179
Critical cyborg literacy, 4–6, 179
    automation and, 27–28, 30–35,
        36, 39, 52
    cyborg bodies and, 59–60, 75–79
    cyborg culture and, 85, 87, 91–92,
        106–107, 110

"Cyborg Manifesto" (Haraway) and,
    114, 119–120, 123–124, 129–
    130, 133–135
    on cyborgs as trouble, 137–141,
        144–146
    on race, 147–155
Critical race theory, 154, 160, 179
Cubacub, Sky, 110–111
Cyberfeminism, 104–111, 179. *See
    also* VNS Matrix
Cyberfeminism Index (Seu), 105–106
Cyberfeminist art, 104–110
"Cyberfeminist Manifesto for the
    21st Century, The" (VNS Matrix),
    104–105
Cybernetics, 9–12, 115–116, 180
Cyborg bodies. *See also* Disability
    biotech enhancement of workers
        and, 47–49
    "Cyborg Manifesto" (Haraway) on,
        124–128
    gestational technologies and, 13,
        58–59, 60, 73–79, 173
    in popular culture, 55–58
Cyborg culture
    Afrofuturism and, 16–17, 101–
        104, 105–106, 153–154, 179
    bioart and, 92–96
    cyberfeminist art and, 104–110
    electronic music and, 13, 99–103
    high fashion and, 96–99, 108,
        110–111
    history of, 85–91
    identity and, 91–92
    key features of, 81–85, 111–112
Cyborg Foundation, 93–94
Cyborg Man (Leeson), 107
"Cyborg Manifesto" (Haraway). *See*
    "Manifesto for Cyborgs: Science,

Technology, and Socialist
Feminism in the 1980s, A"
(Haraway)
Cyborgs, 180. *See also* Automation
(cyborg labor); Cyborg bodies;
Cyborg culture
future of, 19, 165–169
gender and sexuality norms and,
1–4, 18, 35–37
origins of, 9–12, 16, 115–116, 138,
146–147
Cyborgs (Bul), 108–109
Cyborgs (Leeson), 107–108
"Cyborgs and Space" (Clynes and
Kline), 9–10, 115–116
Cyborgs W1–W4 (Lee), 109–110
Cyborg theory, 180. *See also* Critical
cyborg literacy; "Manifesto for
Cyborgs: Science, Technology,
and Socialist Feminism in the
1980s, A" (Haraway)
feminism and, 2–4, 12–16
problems with, 146–147
Cybotron, 99

Da Rimini, Francesca, 104–105
"Data Rituals in Intimate
Infrastructures" (Forlano), 56,
57
"Dawn of the Tryborg" (Weise),
68–69
DeCook, Julia R., 77, 149–150
Deep Sea Dweller (Drexciya), 102–
103
Delany, Samuel L., 103
Deliveroo, 45
Design, 86–88
Design / Not Design (Kawakubo),
98–99

Dewey-Hagborg, Heather, 95
did not feel low, was sleeping
(exhibition), 111
Difference Machines (exhibition),
91–92
Disability
body politics of cripborgs and
tryborgs and, 60, 66–71
"Cyborg Manifesto" (Haraway) and,
126–127
cyborgs as trouble and, 144–146,
155–162
experiences of, 56, 57–58
feminism and, 58–60
future of cyborgs and, 170–171
intimate infrastructures and, 60,
62–66, 68
origins of cyborgs and, 11–12
race and, 75–79
Disability dongles, 158–159
Disability justice, 110–111, 157,
160–161, 180
Disabled cyborgs, 180. *See also*
Disability
Discreet Music (Eno), 13
DJ Spooky, 103
Domestic labor, 35–37
Dourish, Paul, 91
Drexciya, 99, 102–103
Duffy, Brooke Erin, 46
Dunbar-Hester, Christina, 149
Dynamo, 50

Ecofeminism, 15
Economic justice, 49–52
Ekbia, Hamid, 184n7
Electronic Frontier Foundation, 17
Electronic music, 13, 99–103
ENIAC computer, 120

Eno, Brian, 13
Eshun, Kodwo, 101–102
Eubanks, Virginia, 149
Eurocentrism, 82

Faber, Liz, 36
Facebook, 40
FairBnb, 51
Fashion, 96–99, 108, 110–111
Feminism, 180
  cyborg theory and, 2–4, 12–16
    (see also Critical cyborg literacy;
    "Manifesto for Cyborgs: Science,
    Technology, and Socialist
    Feminism in the 1980s, A"
    [Haraway])
  disability and, 58–60
  gestational technologies and, 13,
    58–59, 73–75
  intersectionality and, 170
Fight for $15 campaign, 50
Fordism, 122–123
Forlano, Laura, 8, 56, 62–66, 110–
  111
Freire, Paulo, 152–153

Gender and sexuality norms, 1–4,
  18, 35–37, 120–124, 125–126.
  See also Patriarchy
Gestational labor, 74–75, 180
Gestational technologies, 13, 58–59,
  60, 72–79, 173
Ghost in the Shell (film), 86
Ghost Work (Gray and Suri), 46
Gig economy, 45–46, 173
Gilliam, Terry, 45
Glabau, Danya, 8–9
Glitches, 63–66, 171–172, 173–174,
  181

Glitch feminism, 105–106
Glitch Feminism (Russell), 64,
  81–82
Global village, 89
Good Design (exhibition), 35–37
Google, 40–41
Gordon, Richard E., 121–122
Gray, Mary L., 46
Gregg, Melissa, 184n7
Gropius, Walter, 86
GrubHub, 45

Haagaard, Alex, 158–159
Habeas Viscous (Weheliye),
  150–151
Hadid, Zaha, 90
Hanson Robotics. See Sophia
  (android)
Haraway, Donna, 182. See also
  "Manifesto for Cyborgs: Science,
  Technology, and Socialist
  Feminism in the 1980s, A"
  (Haraway)
Harbisson, Neil, 92–94
Harding, Sandra, 131
Heteromation (Ekbia and Nardi),
  184n7
Hickman, Louis, 159
Hicks, Mar, 120, 149
Hip-hop, 100–101
Home births, 72–73
Homework economy, 121–123
hooks, bell, 152–153
Hopkinson, Nalo, 103
Humanoid robots (androids), 2, 169.
  See also Sophia (android)
Humans (television series), 86
Human Use of Human Beings, The
  (Wiener), 10–11, 115–116

Implantable cardioverter defibrillators (ICDs), 64–65
Infrastructure, 172–173, 181. *See also* Intimate infrastructures
Internet Dream (Paik), 88–89
Intersectional feminism, 170
Intimate infrastructures, 60, 62–66, 68
Ironman (film), 86

Jackson, Liz, 158
Jemisin, N. K., 103
Johnson, Jasmine, 154
Just-in-time scheduling, 38, 51

Kafer, Alison, 138, 155–157, 159–160
Kapoor, Anish, 90
Kawakubo, Rei, 96–99, 108
Keller, Evelyn Fox, 131
Kennedy, Jenny, 36
Kitchen, The (New York), 81–82
Kline, Nathan S., 9–10, 115–116
Knee Cap (Leeson), 107
Kraftwerk, 99–100
Kronos, 51
Kurzweil, Ray, 17

Labaco, Ron, 90
Labor, 181. *See also* Automation (cyborg labor); Gestational labor
disability and, 69–70
future of, 173–174
racial and class inequalities and, 151–152
women and, 119–124
Lang, Fritz, 86
Language, 175
Latour, Bruno, 37

Lee, Suzanne, 95
Leeson, Lynn Hershman, 106–108
Lewis, Sophie, 74–75
Light, Jennifer, 120
Liu, Ani, 95
Lovecraft Country (television series), 147–148
Lovelace, Ada, 107–108
Ludd, Ned, 37–38
Luddites, 37–38

Ma (gap), 98
Mäki-Reinikka, Kasperi, 92, 93, 94
"Manifesto for Cyborgs: Science, Technology, and Socialist Feminism in the 1980s, A" (Haraway)
on bodies, 124–128
on cyborg as metaphor for personhood and politics, 115–119
on cyborgs as trouble, 137–140, 176–177
feminist theory and, 128–133
as foundational text, 13–14, 113–114, 133–135, 163, 169
on labor, 119–124
responses and critiques of, 14–17, 110, 126–127, 138, 149–150, 152, 155
VNS Matrix and, 16–17, 104–105
Man-Machine (Kraftwerk), 99–100
Martin, Emily, 131
Marx, Karl, 181
McKittrick, Katherine, 102–103
McLuhan, Marshall, 89
Mead, Margaret, 11
mEat me (Schubert), 95

Mechanical Turk platform, 30, 45, 46, 50–51
Medicaid, 70
Medicare, 70
Merce/Digital (Paik), 88
Metaverse, 140, 142–143, 144–145
#MeToo movement, 50
Metropolis (film), 86
Midwifery, 72–73
Militarism, 16, 115–117, 120, 131–133, 138, 173
Minimum wage, 50
Monae, Janelle, 103
More Brilliant than the Sun (Eshun), 101–102
More Than Machines? (Voss), 44
More Work for Mother (Cowan), 36
Mu (negation, emptiness, and nothingness), 98
Mullins, Amy, 69
Mundane Afrofuturist Manifesto, The (Syms), 103–104
Music, 13, 99–101
Musk, Elon, 138, 140–142, 144–145

Nardi, Bonnie, 184n7
Nature, 15, 77–79, 150–151, 169, 181
Nature versus nurture debate, 117–118, 124–125, 127, 131–133
Nelson, Alondra, 79
Nelson, Mallory Kay, 66–68, 70
Neuralink, 140–142, 144–145
New York Times (newspaper), 43, 46–47, 68–69, 100
Noble, Safia, 149
(Not) Getting Paid to Do What You Love (Duffy), 46

Object/Subject (Kawakubo), 99
Oram, Daphne, 100
Orphan Black (television series), 86
Orr, Julian, 184n7
Ortner, Sherry, 131
Osrow, Harold, 35–37
Oudshoorn, Nelly, 64–66
Out of Hand: Materializing the Post Digital (exhibition), 90–91

Pacemakers, 64, 65–66
Paik, Nam June, 88–89, 106–107
Patriarchy, 117–118, 128–129, 131
Physical labor, 42
Pierce, Julianne, 104–105
Planet Rock, 101
Plant, Sadie, 104–105, 179
Politics, 176
Post-cyberfeminism, 105–106
Posthumanism, 146, 149, 181
Praxis, 152–153

Quik-Suds Semi-Automatic Dishwasher, 35–37

Ra, Sun, 103
Race and racism, 2, 14, 75–79, 127–128, 147–155, 160–162
Radio-frequency identification chips, 48
Rebirth Garments, 110–111
Rebirth of a Nation (DJ Spooky), 103
Remote and distributed work, 41–45
RE-WIRED/RE-MIXED: Event for Dismembered Body (Stelarc), 94–95
Rhizome, 105–106
Ribas, Moon, 93–94

Riley, Boots, 48–49
Rivera, Alex, 41
"Robot, The" (dance move), 100–101
Robot K-456, 88
Robots, 86, 182. *See also* Cyborgs
Rockefeller Foundation, 11
Rosenblat, Alex, 46
Rosenthal, Stephanie, 109–110
Rovner, Lisa, 100
R.U.R. (Čapek), 86
Russell, Legacy, 64, 81–82, 91
Ryan, Jeri, 57
Ryan, Tina Rivers, 91–92

Sandoval, Chela, 15–16
Schalk, Sami, 79
Schubert, Theresa, 95
Schütte, Uwe, 100
Science fiction, 85–86, 103, 142
Scientific management (Taylorism), 38
Second Sight, 145–146
Second-wave feminism, 13, 130–131
Self-checkout, 39–40
Self/Other (Kawakubo), 99
Seu, Mindy, 105–106
Sexual harassment, 50
Shannon, Claude, 11
Shapiro, Peter, 100
Sharpe, Christina, 77–79
Shew, Ashley, 11–12, 66–68, 70, 170–171
Sinnreich, Aram, 100–101
Sisters with Transistors (documentary film), 100
Situatedness, 146, 167, 169–170, 177, 182

Slavery, 48–49
Sleep Dealer (film), 31, 41
Smart Wife, The (Strengers and Kennedy), 36
Social justice, 49–52
Sophia (android), 1–3, 18, 81
Sorry to Bother You (film), 31, 48–49
Soulsonic Force, 101
Spiegel, Laurie, 100
Splan, Laura, 95
Star, Susan Leigh, 16, 181
Starrs, Josephine, 104–105
Star Trek: Voyager (television series), 55–57
Stelarc, 94–95
Stevens, Bethany, 66–68, 70
Strengers, Yolande, 36
Stuff of Bits, The (Dourish), 91
Suchman, Lucy, 184n7
Suri, Siddarth, 46
Surrogacy, 74–75
Syms, Martine, 103–104

Tales of Hoffmann, The (opera), 85–86
Talking about Machines (Orr), 184n7
Taylor, Frederick Winslow, 38
Taylorism (scientific management), 38
Technoableism, 156–157
Technochauvinism, 18–19
Technodeterminism, 182
    automation and, 38–39, 47
    cyborg culture and, 82, 87, 89, 90
    cyborgs as trouble and, 149, 160
    disability and, 144–145

Technology, 182. *See also* Cyborg
    theory
  domestic labor and, 35–37
  histories and contexts of, 17–19
  women and, 119–121
Techno-optimism, 42, 138–146, 156
Technopessimism, 149
Technoutopianism, 149
Textiles, 96–99, 108, 110–111
Third-wave feminism, 13
"Third World" feminisms, 130
Transhumanism, 17, 48–49, 106–
    107, 138–146, 160, 177, 182
"Transmobility: Rethinking
    the Possibilities in Cyborg
    (Cripborg) Bodies" (Nelson,
    Shew, and Stevens), 66–68, 70
Tryborgs, 60, 68–71
Turkopticon, 50
Tyre, Michael, Sr., 43

Uber, 45
Uberland (Rosenblat), 46
Uncle (Paik), 88

Vanouse, Paul, 91–92
VNS Matrix, 16–17, 104–105, 106,
    108
Vora, Kalindi, 151–152
Voss, Laura, 44

Wage inequality, 50
Weaver, Warren, 11
Weheliye, Alexander, 150–151
Weise, Jillian, 68–69, 126–127,
    170–171
Wenzel, Kaethe, 95
"[White] Cyborg's Manifesto, A"
    (DeCook), 149–150

Wiener, Norbert, 10–11, 115–116
Williams, Rua, 158–159
Williamson, Bess, 70–71
Workers Lab, 51

Xenofeminism, 105–106

Zen philosophy, 96–99
Zero Theorem, The (film), 31, 45
Zuckerberg, Mark, 138, 140–141,
    142–143, 144–145

LAURA FORLANO is a Fulbright award-winning writer, social scientist, and design researcher. She is Professor in the College of Arts, Media and Design at Northeastern University. Forlano's research is focused on design, computation, and disability with an emphasis on creative practice. She is coeditor of three books: *Bauhaus Futures* (MIT Press, 2019), *digitalSTS* (Princeton University Press, 2019), and *From Social Butterfly to Engaged Citizen* (MIT Press, 2011). Forlano received her PhD in communications from Columbia University.

DANYA GLABAU is an anthropologist and STS scholar as well as Industry Assistant Professor in the Department of Technology, Culture and Society and Director of the Science and Technology Studies program at the Tandon School of Engineering at New York University. Glabau's work is centered in feminist STS and medical anthropology, and looks at topics including food allergies, medical activism, cyborgs and feminist cybercultures, the political economy of the global pharmaceutical industry, and the risks posed by digital health. She received her PhD from the Department of Science and Technology Studies at Cornell University. Her first scholarly book, *Food Allergy Advocacy: Parenting and the Politics of Care*, examines how food allergy activism in the United States is shaped by reproductive politics, particularly norms of whiteness and femininity in caretaking and the ideal of the nuclear family.